Precalculus: With Limits

Precalculus: With Limits

Editor

Nik Pachis

scitus academics

Precalculus: With Limits
Edited by **Nik Pachis**

ISBN: 978-1-68117-450-1
Library of Congress Control Number: 2016937609

www.scitusacademics.com
Box No. 4766, 616 Corporate Way,
Suite 2, Valley Cottage,
NY 10989

Printed in United States of America on Acid Free Paper

Preface

Calculus is a conceptual framework which provides systematic techniques for solving problems. These problems are appropriately applicable to analytic geometry and algebra. Therefore precalculus gives you the background for the mathematical concepts, problems, issues and techniques that appear in calculus, including trigonometry, functions, complex numbers, vectors, matrices, and others. Precalculus prepares students for calculus somewhat differently from the way that pre-algebra prepares students for algebra. While pre-algebra often has extensive coverage of basic algebraic concepts, precalculus courses might see only small amounts of calculus concepts, if at all, and often involves covering algebraic topics that might not have been given attention in earlier algebra courses. The book Precalculus: With Limits includes a brief algebra review to the core precalculus topics along with coverage of analytic geometry in three dimensions and an introduction to concepts covered in calculus. The book is designed for a variety of students with different mathematical needs. For those students who will take additional mathematics, the book will provide the proper foundation of skills, understanding, and insights necessary for success in further courses. For those students who will not pursue further mathematics, the extensive emphasis on applications and modeling will demonstrate the usefulness and applicability of mathematics in the world today. Many of the applied problems in this text are actually real problems that people have had to solve on the job.

Table of Contents

CHAPTER 1

On Some Fundamental Integrodifferential Inequalities

Zareen A. Khan

Department of Mathematics, Princess Noura Bint Abdurehman University, Riyadh, KSA

ABSTRACT

The aim of this present paper is to establish some new integrodifferential inequalities of Gronwall type involving functions of one independent variable which provide explicit bounds on unknown functions. The inequalities given here can be used in the analysis of a class of differential equations as handy tools.

KEYWORDS

Integral Inequalities, Two Independent Variables, Nondecreasing, Nonincreasing

1. INTRODUCTION

The differential and integral inequalities occupy a very privileged position in the theory of differential and integral equations. In recent years, these inequalities have been greatly enriched by the recognition of their potential and intrinsic worth in many applications of the applied sciences. The integrodifferential inequalities recently established by Gronwall and others [1] -[12] have attracted considerable attention in the theory of differential and integral equations. This fact encourages us to find the explicit bounds on some fundamental integrodifferential inequalities which can be applied fairly well to achieve a diversity of desired goals. In [3] , Pachpatte (1977) gave the following useful integrodifferential inequality:

Let $u(t)$, $\dot{u}(t)$ and $b(t)$ be nonnegative continuous functions defined on R_+ and $a > 0$ is constant. If

$$\dot{u}(t) \leq a + \int_0^t u(s)\dot{u}(s)\left(u(s)+\dot{u}(s)\right)ds \tag{1.1}$$

for $t \in R_+$ and $E(t)$ is defined by

$$E(t) = 1 - \left[a + u(0)\right]\int_0^t e^{\sigma} b(\sigma)d\sigma$$

then

$$\dot{u}(t) \leq a\exp\left(\left[a + u(0)\right]\int_0^t e^{s} b(s)\left(E(s)\right)^{-1}ds\right)$$

for $t \in [0, \beta)$ where

$$\beta = \sup\{t \in R_+ : E(t) > 0\}.$$

Our goal in this paper is to establish new explicit bounds on some basic integrodifferential inequalities of one independent variable which will be equally important in handling the inequality (1.1). Given application in this paper also illustrates the usefulness of our result.

2. MAIN RESULTS

Theorem 2.1: Let $u(t)$, $\dot{u}(t)$ and $k(t)$ be nonnegative continuous functions defined on $I = [0,\infty)$ for which the inequality

$$\dot{u}(t) \le c + \int_0^t k(s)\dot{u}(s)\left(\dot{u}^p(s) + u^2(s)\right)ds, \quad t \in I \tag{2.1}$$

holds, where c is positive constant and $p \ge 1$. If

$$1 - p^2(2p-1)\left[c^{p^2} + u(0) + u^{2p}(0)\right]^{2p-1} \int_0^t k(s)e^{(2p-1)s}ds > 0, \quad t \in I \tag{2.2}$$

and

$$1 - \left[c^{p^2} + u^2(0)\right]^p \int_0^t \left(p^2k(s) - 2p\right)\exp\left(2p\int_0^s Q_1(r)dr\right)ds > 0, \quad t \in I \tag{2.3}$$

then

$$\dot{u}(t) \le c\exp\left[\int_0^t k(s)R_1(s)ds\right], \tag{2.4}$$

$\forall t \in I$, and

$$R_1(t) = \frac{\left[c^p + u^2(0)\right]\exp\left(2p\int_0^t Q_1(s)\mathrm{d}s\right)^{\frac{1}{p}}}{\left[1-\left[c^p + u^2(0)\right]^p \int_0^t \left(p^2k(s)-2p\right)\exp\left(2p\int_0^s Q_1(r)\mathrm{d}r\right)\mathrm{d}s\right]^{\frac{1}{p}}} \tag{2.5}$$

also

$$Q_1(t) = \frac{\left[c^{p^2} + u^{2p}(0) + u(0)\right]\mathrm{e}^t}{\left[1-p^2(2p-1)\left[c^{p^2} + u^{2p}(0) + u(0)\right]^{2p-1}\int_0^y k(s)\mathrm{e}^{(2p-1)}\mathrm{d}s\right]^{\frac{1}{2p-1}}} \tag{2.6}$$

Proof: Define a function $m(t)$ by the right-hand side of (2.1). Then

$$m(t) = c + \int_0^t k(s)\dot{u}(s)\left(\dot{u}^p(s) + u^2(s)\right)\mathrm{d}s \tag{2.7}$$

where

$$m(0) = c \tag{2.8}$$

Then from (2.1) and (2.7), we have

$$\dot{u}(t) \le m(t) \tag{2.9}$$

Integrating both sides of (2.9) from 0 to t, we observe that

$$u(t) \le u(0) + \int_0^t m(s)\,\mathrm{d}s \tag{2.10}$$

Differentiating both sides of (2.7) with respect to t and using (2.9) and (2.10), we get

$$\dot{m}(t) \le k(t)m(t)\left[m^p(t) + \left[u(0) + \int_0^t m(s)\,\mathrm{d}s\right]^2\right] \tag{2.11}$$

Define a function $n(t)$ by the right-hand side of (2.11), then

$$n(t) = m^p(t) + \left[u(0) + \int_0^t m(s)\,\mathrm{d}s\right]^2 \tag{2.12}$$

Where

$$n(0) = c^p + u^2(0) \tag{2.13}$$

It is clear that

$$m(t) \le n(t) \tag{2.14}$$

By using (2.12) in (2.11), we have

$$\dot{m}(t) \le k(t)m(t)n(t) \tag{2.15}$$

Differentiating both sides of (2.12) with respect to t, we get

$$\dot{n}(t) = pm^{p-1}(t)\dot{m}(t) + 2\left[u(0) + \int_0^t m(s)\,\mathrm{d}s\right]m(t)$$

By using (2.14) and (2.15) in the above equation, we observe that

$$\dot{n}(t) \le n^{p+1}(t)\left[pk(t)-2\right]+2n(t)\left[n^{p}(t)+u(0)+\int_0^t n(s)\mathrm{d}s\right] \tag{2.16}$$

Let

$$z(t)=n^{p}(t)+u(0)+\int_0^t n(s)\mathrm{d}s \tag{2.17}$$

Where

$$z(0)=n^{p}(0)+u(0)=\left[c^{p}+u^{2}(0)\right]^{p}+u(0) \tag{2.18}$$

And

$$n(t)\le z(t) \tag{2.19}$$

Using (2.17) in (2.16), we get

$$\dot{n}(t)\le n^{p+1}(t)\left[pk(t)-2\right]+2n(t)z(t) \tag{2.20}$$

Differentiating both sides of (2.17) with respect to t, we get

$$\dot{z}(t)=pn^{p-1}(t)\dot{n}(t)+n(t) \tag{2.21}$$

Inequality (2.21) by using (2.19) and (2.20), and since $z^{p+1}(t)\le z^{2p}(t)$ if $p\ge 1$ takes the form

$$z^{-2p}(t)\dot{z}(t)-z^{-2p+1}(t)\le p^{2}k(t) \tag{2.22}$$

Let

$$r(t)=z^{-2p+1}(t) \tag{2.23}$$

Where

$$r(0)=z^{-2p+1}(0)=\left[\left[c^{p}+u^{2}(0)\right]^{p}+u(0)\right]^{-2p+1} \tag{2.24}$$

Differentiating both sides of (2.23) with respect to t, we get

$$\dot{r}(t)=(-2p+1)z^{-2p}(t)\dot{z}(t) \tag{2.25}$$

Inequality (2.22) by using (2.23) and (2.25), takes the form

$$\dot{r}(t)+(2p-1)r(t)\geq -p^{2}(2p-1)k(t) \tag{2.26}$$

Multiplying both sides of (2.26) by $e^{(2p-1)t}$ and integrating the resulting inequality from 0 to t, and using (2.24), we have

$$r(t)\geq\left[c^{p^{2}}+u^{2p}(0)+u(0)\right]^{-(2p-1)}\mathrm{e}^{-(2p-1)t}-p^{2}(2p-1)\mathrm{e}^{-(2p-1)t}\int_{0}^{t}k(s)\mathrm{e}^{(2p-1)s}\mathrm{d}s$$

By using (2.23) in the above inequality, it can be seen that

$$z(t)\leq\left[\left[c^{p^{2}}+u^{2p}(0)+u(0)\right]^{-(2p-1)}\mathrm{e}^{-(2p-1)t}\right]^{-\frac{1}{2p-1}}$$

$$\times\left[1-p^{2}(2p-1)\mathrm{e}^{-(2p-1)t}\left[c^{p^{2}}+u^{2p}(0)+u(0)\right]^{2p-1}\int_{0}^{t}k(s)\mathrm{e}^{(2p-1)}\mathrm{d}s\right]^{-\frac{1}{2p-1}}.$$

which can be rewritten as

$$z(t) \le \frac{\left[c^{p^2} + u^{2p}(0) + u(0)\right]e^t}{\left[1 - p^2(2p-1)\left[c^{p^2} + u^{2p}(0) + u(0)\right]^{2p-1} \int_0^t (s)e^{(2p-1)s}ds\right]^{\frac{1}{2p-1}}} = Q_1(t)$$

(2.27)

Using (2.27) in (2.20), we observe that

$$n^{-(p+1)}(t)\dot{n}(t) - 2Q_1(t)n^{-p}(t) \le \left[pk(t) - 2\right] \tag{2.28}$$

Let

$$h(t) = n^{-p}(t) \tag{2.29}$$

Where

$$h(0) = n^{-p}(0) = \left[c^p + u^2(0)\right]^{-p} \tag{2.30}$$

Differentiating both sides of (2.29) with respect to t, we get

$$\dot{h}(t) = -pn^{-p-1}(t)\dot{h}(t) \tag{2.31}$$

Inequality (2.28) by using (2.29) and (2.31), takes the form

$$\dot{h}(t) + 2pQ_1(t)h(t) \ge -\left[p^2k(t) - 2p\right] \tag{2.32}$$

Multiplying both sides of (2.32) by $\exp\left(2p\int_0^t Q(T)dT\right)$ and integrating the resulting inequality from 0 to t, and using (2.29) and (2.30), we have

$$n^{-p}(t) \geq \left[c^p + u^2(0)\right]^{-p} \exp\left(-2p\int_0^t Q_1(s)\mathrm{d}s\right) - \exp\left(-2p\int_0^t Q_1(s)\mathrm{d}s\right) \times \int_0^t \left(p^2 k(s) - 2p\right)\exp\left(2p\int_0^s Q_1(\alpha)\mathrm{d}\alpha\right)\mathrm{d}t$$

which can be rewritten as

$$n(t) \leq \frac{\left[c^p + u^2(0)\right]\exp\left(2p\int_0^t Q(s)\mathrm{d}s\right)^{\frac{1}{p}}}{\left[1 - \left[c^p + u^2(0)\right]^p \int_0^t \left[p^2 k(s) - 2p\right]\exp\left(2p\int Q(\alpha)\mathrm{d}\alpha\right)\mathrm{d}t\right]^{\frac{1}{p}}} = R_1(t)$$

(2.33)

From (2.15) and (2.33), we get

$$\frac{\dot{m}(t)}{m(t)} \leq k(t)R_1(t)$$

Integrating both sides of the above inequality from 0 to t, and from (2.8), we observe that

$$m(t) \leq c\exp\left[\int_0^t k(s)R_1(s)\mathrm{d}s\right] \tag{2.34}$$

From (2.9) and (2.34), we have

$$\dot{u}(t) \leq c\exp\left[\int_0^t k(s)R_1(s)\mathrm{d}s\right]$$

Application: As an application we obtain the bound on the solution of the differential equation of the formulation of the form

$$\ddot{x}(t) = \dot{x}(t) f\left(t, \dot{x}^p, x^2\right) \tag{2.35}$$

with the given initial conditions

$$\dot{x}(0)=a, \quad x(0)=c \tag{2.36}$$

where $f: I\times R\times R\rightarrow R$ is a continuous function and a,c are real constants. $I=[t_0,\infty)$, $t_0\geq 0$. Here we assume that the solution $x(t)$ of (2.35) and (2.36) exists on R_+ Assume that the function in (2.35) satisfies the condition

$$\left|f\left(t,\dot{x}^2,x\right)\right|\leq k(t)\left[\left|\dot{x}^p\right|+x^2\right], \quad \forall t\in I \tag{2.37}$$

where $k(t)$ is a real valued nonnegative continuous function defined on I. If

$$1-p^2(2p-1)\left[a^{p^2}+c+c^{2p}\right]^{2p-1}\int_0^t k(s)e^{(2p-1)s}ds>0, \quad t\in I$$

And

$$1-\left[a^{p^2}+c^2\right]^p\int_0^t\left(p^2k(s)-2p\right)\exp\left(2p\int_0^s Q_1(r)dr\right)ds>0, \quad t\in I$$

then the bounds on the solution (2.35) takes the form

$$\left|\dot{x}(t)\right|\leq a\exp\left(\int_0^t k(s)R(s)ds\right) \tag{2.38}$$

$\forall t\in I$, where, $p\geq 1$, and

$$R(t)=\frac{\left[a^{p}+c^{2}\right]\exp\left(2p\int_{0}^{t}Q_{1}(s)\mathrm{d}s\right)^{\frac{1}{p}}}{\left[1-\left[a^{p}+c^{2}\right]^{p}\int_{0}^{t}\left(p^{2}k(s)-2p\right)\exp\left(2p\int_{0}^{s}Q_{1}(r)\mathrm{d}r\right)\mathrm{d}s\right]^{\frac{1}{p}}}$$

Also

$$Q_{1}(t)=\frac{\left[a^{p^{2}}+c^{2p}+c\right]\mathrm{e}^{t}}{\left[1-p^{2}(2p-1)\left[a^{p^{2}}+c^{2p}+c\right]^{2p-1}\int_{0}^{y}k(s)\mathrm{e}^{(2p-1)}\mathrm{d}s\right]^{\frac{1}{2p-1}}}$$

Proof: Integrating both sides of (2.35) from 0 to t, and using (2.36), we observe that

$$\dot{x}(t)=a+\int_{0}^{t}\dot{x}(s)f\left(s,\dot{x}^{p}(s),x^{2}(s)\right)\mathrm{d}s$$

Taking absolute values of both sides of the above equation and using (2.37), we get

$$\left|\dot{x}(t)\right|\le a+\int_{0}^{t}k(s)\left|\dot{x}(s)\right|\left[\left|\dot{x}^{p}(s)\right|+x^{2}(s)\right]\mathrm{d}s \tag{2.39}$$

The remaining proof is the same as Theorem 2.1 by following the same steps from (2.7)-(2.35) in (2.39) with suitable modifications, we get the required bound of (2.35).

We note that many generalizations, extensions, variants and applications of the inequality given in this paper are possible and we hope that the result given here will assure greater importance in near future.

REFERENCES

1. Ragab, A.A. (2000) On Integral Inequalities and Their Applications. Kyungpook Mathematical Journal, 40, 357-370.

2. Pachpatte, B.G. (2001) On Some Fundamental Integral Inequalities and Their Discrete Analogues. Journal of Inequalities in Pure and Applied Mathematics, 2, 1-13.

3. Pachpatte, B.G. (1977) On Some Fundamental Integrodifferential and Integral Inequalities. Analele. Stiintifice ale. Universitati.Al.I.Cuza.din.Iasi, 33, 77-86.

4. Pachpatte, B.G. (1973) A Note on Gronwall-Bellman Inequalities. Journal of Mathematical Analysis and Applications, 44, 758-762.

5. Pachpatte, B.G. (1975) On Some Integral Inequalities Similar to Bellman-Bihari Inequality. Journal of Mathematical Analysis and Applications, 49, 794-802.

6. Pachpatte, B.G. (1974) An Integral Inequality Similar to Bellman-Bihari Inequality. Bulletin of the Greek Mathematical Society, 15, 7-12.

7. Pachpatte, B.G. (1977) A Note on Gronwall Type Integral and Integrodifferential Inequalities. Tamkang Journal of Mathematics, 8, 53-59.

8. Ullrich, D.F. (1969) Boundary Value Problems for a Class of Nonlinear Second Order Differential Equations. Journal of Mathematical Analysis and Applications, 28, 188-210.

9. Beckenbach, E.F. and Bellman, R. (1961) Inequalities. Springer-Verlag, Berlin Heidelberg.

10. Meng, F.M. and Li, W.N. (2004) On Some Integral Inequalities and Their Applications. Applied Mathematics and Computation, 148, 381-392.

11. Hooker, J.W. (1969) Existence and Oscillations Theorems for a Class of Nonlinear Second Order Differential Equations. Journal of Differential Equations, 5, 283-306.

12. Ma, Q.H. and Debnath, L. (2003) A More Generalized Gronwall-Like Integral Inequality with Applications. International Journal of Mathematics and Mathematical Sciences, No. 15, 927-934.

CHAPTER 2

On a Class of Gronwall-Bellman Type Inequalities

Zareen A. Khan

Department of Mathematics, Princess Noura Bint Abdurehman University, Riyadh, KSA

ABSTRACT

The aim of the present paper is to establish some new integral inequalities of Gronwall type involving functions of two independent variables which provide explicit bounds on unknown functions. The inequalities given here can be used as tools in the qualitative theory of certain partial differential and integral equations.

KEYWORDS

Integral Inequalities, Two Independent Variables, Partial Differential Equations, Nondecreasing, Nonincreasing

1. INTRODUCTION

The Gronwall type integral inequalities provide a necessary tool for the study of the theory of differential equations, integral equations and inequalities of the various types (please, see Gronwall[1] and Guiliano [2]). Some applications of this result to the study of stability of the solution of linear and nonlinear differential equations may be found in Bellman [3] . Numerous applications to the existence and uniqueness theory of differential equations may be found in Nemyckii-Stepanov[4] , Bihari [5] , and Langenhop [6] . During the past few years several authors (see references below and some of the references cited therein) have established several Gronwall type integral inequalities in one or two independent real variables [1] -[14] . Of course, such results have application in the theory of partial differential equations and Volterra integral equations.

In [14] , Pachpatte investigated the following inequality:

Lemma: Let u, f and g be nonnegative continuous functions defined on R_+ and c be nonnegative constant. Let $w(t,r)$ be a nonnegative continuous function defined for $t \in R_+, 0 \le r < \infty$, and monotonic nondecreasing with respect to t for ant fixed $t \in R_+$. If

$$u^2(t) \le c^2 + 2\int_0^t u(s)w(s,u(s))\mathrm{d}s$$

For $t \in R_+$, then

$$u(t) \le r(t), \quad t \in R_+$$

where $r(t)$ is the maximal solution of

$$r'(t) = w(t,r(t)), \quad r(0) = c$$

For $t \in R_+$.

2. MAIN RESULTS

Theorem 2.1: Let $u(x,y)$ and $a(x,y)$ be nonnegative continuous functions defined on $x,y \in R_+$. Let $w(x,y)$ be a positive continuous and nondecreasing functions in both variables and defined for $x,y \in R_+$. If

$$u(x,y) \le w(x,y) + \int_0^x\int_0^y u(s,t)\mathrm{d}t\mathrm{d}s + \int_0^x\int_0^y a(s,t)u^p(s,t)\mathrm{d}t\mathrm{d}s \tag{2.1}$$

Then

$$u(x,y) \le E(x,y)\left[w(x,y)+r(x,y)\right], \quad \forall t \in R_+$$

where $0 < p < 1 \cdot p + q = 1, q > 0$ and

$$E(x,y) = \left[1 + (1-p)\int_0^x\int_0^y a(s,t)m^{-q}(s,t)\,dtds\right]^{\frac{1}{1-p}}, \tag{2.2}$$

and $r(x,y)$ is the maximal solution of

$$r_{xy}(x,y) = E(s,t)\left[w(s,t)+r(s,t)\right], \; r_x(x,0)=0, \; r(0,y)=0 \tag{2.3}$$

Proof: Define a function $m(x,y)$ by the right-hand side of (2.1).Then

$$m(x,y) = w(x,y) + \int_0^x\int_0^y u(s,t)\,dtds \tag{2.4}$$

By using (2.4) in (2.1), we get

$$u(x,y) \le m(x,y) + \int_0^x\int_0^y a(s,t)u^p(s,t)\,dtds \tag{2.5}$$

Since $m(x,y)$ is a positive continuous and nondecreasing function, then

$$\frac{u(x,y)}{m(x,y)} \le 1 + \int_0^x\int_0^y a(s,t)m^{-q}(s,t)\left[\frac{u(s,t)}{m(s,t)}\right]^p dtds \tag{2.6}$$

Let

$$\frac{u(x,y)}{m(x,y)} \le v(x,y) \tag{2.7}$$

From (2.6) and (2.7), we observe that

$$v(x,y) = 1 + \int_0^x \int_0^y a(s,t) m^{-q}(s,t) v^p(s,t) \, dt ds \tag{2.8}$$

And

$$v(0,y) = 1, \quad v(x,0) = 1, \quad \text{and } v_x(x,0) = 0, \; v_y(0,y) = 0 \tag{2.9}$$

Differentiating both sides of (2.8) with respect to x and y, we get

$$v_{xy}(x,y) \le a(x,y) m^{-q}(x,y) v^p(x,y)$$

$$\frac{\partial}{\partial y}\left[\frac{v_x(x,y)}{v^p(x,y)}\right] \le a(x,y) m^{-q}(x,y) \tag{2.10}$$

By keeping first x fixed in (2.10) and set $y = t$ and integrate from 0 to y then again keeping y fixed, set $x = s$ and integrate from 0 to x respectively and using (2.9), we get

$$v(x,y) \le \left[1 + (1-p) \int_0^x \int_0^y a(s,t) m^{-q}(s,t) \, dt ds\right]^{\frac{1}{1-p}} = E(x,y) \tag{2.11}$$

From (2.7) and (2.11), it is clear that

$$u(x,y) \le E(x,y)m(x,y)$$

From (2.4), it can be restated as

$$u(x,y) \le E(x,y)\left[w(x,y)+r(x,y)\right],$$

where $r(x,y)$ is the maximal solution of

$$r_{xy}(x,y) = E(s,t)\left[w(s,t)+r(s,t)\right], \quad r_x(x,0)=0, \; r(0,y)=0$$

This completes the proof.

Theorem 2.2: Let $u(x,y), a(x,y), w(x,y)$ be defined as in Theorem 2.1. If

$$u(x,y) \le w(x,y) + \int_0^x \int_0^y u^p(s,t)\,dt\,ds + \int_0^x \int_0^y a(s,t)u^p(s,t)\,dt\,ds \qquad (2.12)$$

Then

$$u(x,y) \le E(x,y)\left[w(x,y)+r(x,y)\right], \quad \forall t \in R_+$$

where $0 < p < 1 \cdot p + q = 1, q > 0$ and

$$E(x,y) = \left[1 + (1-p)\int_0^x \int_0^y a(s,t)m^{-q}(s,t)\,dt\,ds\right]^{\frac{1}{1-p}}, \qquad (2.13)$$

where $r(x,y)$ is the maximal solution of

$$r_{xy}(x,y)=\left[E(s,t)\left[w(s,t)+r(s,t)\right]\right]^{p},\ \ r_x(x,0)=0,\ r(0,y)=0$$

Proof: Define a function $m(x,y)$ by the right-hand side of (2.12).Then

$$m(x,y)=w(x,y)+\int_0^x\int_0^y u^p(s,t)\,\mathrm{d}t\mathrm{d}s \tag{2.14}$$

By using (2.14) in (2.12), we get

$$u(x,y)\le m(x,y)+\int_0^x\int_0^y a(s,t)u^p(s,t)\,\mathrm{d}t\mathrm{d}s \tag{2.15}$$

By following the same steps of Theorem 2.1 from (2.5)-(2.11), we get

$$v(x,y)\le\left[1+(1-p)\int_0^x\int_0^y a(s,t)m^{-q}(s,t)\,\mathrm{d}t\mathrm{d}s\right]^{\frac{1}{1-p}}=E(x,y) \tag{2.16}$$

From (2.7), (2.14) and (2.16), we observe that

$$u(x,y)\le E(x,y)\left[w(x,y)+r(x,y)\right],$$

where $r(x,y)$ is the maximal solution of

$$r_{xy}(x,y)=\left[E(s,t)\left[w(s,t)+r(s,t)\right]\right]^{p},\ \ r_x(x,0)=0,\ r(0,y)=0$$

This completes the proof.

Theorem 2.3: Let $u(x,y)$, $a(x,y)$, $w(x,y)$ be defined as in Theorem 2.1. If

$$u^p(x,y) \leq w(x,y) + \int_0^x \int_0^y u(s,t)\,dt\,ds + \int_0^x \int_0^y a(s,t) u^p(s,t)\,dt\,ds \tag{2.17}$$

Then

$$u(x,y) \leq \left[E(x,y)\left[w(x,y) + r(x,y)\right]\right]^{\frac{1}{p}}, \quad \forall t \in R_+$$

where $p > 1.$ and

$$E(x,y) = \exp\left[\int_0^x \int_0^y a(s,t)\,dt\,ds\right], \tag{2.18}$$

where $r(x,y)$ is the maximal solution of

$$r_{xy}(x,y) = E(s,t)\left[w(s,t) + r(s,t)\right], \quad r_x(x,0) = 0, \; r(0,y) = 0$$

Proof: Define a function $z(x,y)$ by the right-hand side of (2.17).Then

$$u^p(x,y) \leq z(x,y) \Rightarrow u(x,y) \leq z^{\frac{1}{p}}(x,y) \tag{2.19}$$

where $$z(x,y)=m(x,y)+\int_0^x\int_0^y a(s,t)u^p(s,t)\,dt\,ds$$

By using (2.19) in the above equation, we get

$$z(x,y)\le m(x,y)+\int_0^x\int_0^y a(s,t)z(s,t)\,dt\,ds \tag{2.20}$$

and

$$m(x,y)=w(x,y)+\int_0^x\int_0^y u(s,t)\,dt\,ds \tag{2.21}$$

Since $m(x,y)$ is a positive continuous and nondecreasing function, then from (2.20),

$$\frac{z(x,y)}{m(x,y)}\le 1+\int_0^x\int_0^y a(s,t)\frac{z(s,t)}{m(s,t)}\,dt\,ds \tag{2.22}$$

Let

$$\frac{z(x,y)}{m(x,y)}\le v(x,y) \tag{2.23}$$

where

$$v(x,y)=1+\int_0^x\int_0^y a(s,t)\frac{z(s,t)}{m(s,t)}\,dt\,ds \tag{2.24}$$

And

$$v(0,y)=1, \quad v(x,0)=1, \quad \text{and } v_x(x,0)=0, \quad v_y(0,y)=0 \tag{2.25}$$

Differentiating both sides of (2.24) with respect to x and y, and from (2.23), we get

$$v_{xy}(x,y) \le a(x,y)v(x,y)$$

$$\frac{\partial}{\partial y}\left[\frac{v_x(x,y)}{v(x,y)}\right] \le a(x,y) \tag{2.26}$$

By keeping first x fixed in (2.26) and set $y=t$ and integrate from 0 to y then again keeping y fixed, set $x=s$ and integrate from 0 to x respectively and using (2.25), we get

$$v(x,y) \le \exp\left[\int_0^x\int_0^y a(s,t)\,dt\,ds\right] = E(x,y) \tag{2.27}$$

From (2.19), (2.23) and (2.27), it is clear that

$$u(x,y) \le \left[E(x,y)\left[w(x,y)+r(x,y)\right]\right]^{\frac{1}{p}},$$

where $r(x,y)$ is the maximal solution of

$$r_{xy}(x,y) = E(s,t)\left[w(s,t)+r(s,t)\right], \quad r_x(x,0)=0, \; r(0,y)=0$$

This completes the proof.

Application: As an application, let us consider the bound on the solution of a nonlinear hyperbolic partial differential equation of the form

$$u_{xy}(x,y) = A\left[x,y,u(s,t),v(s,t)\right] \tag{2.28}$$

with the given boundary conditions

$$u(x,y_0) = a_1(x),\ u(x_0,y) = a_2(y),\ \ a_1(x_0) = a_2(y_0) = 0 \tag{2.29}$$

where $u \in C[D,R],\ A \in C[D \times R,R],$ and

$D = \{(x,y), x \geq 0, y \geq 0\}$ such that

$$\left|A(x,y,u,v)\right| \leq |u| + |v| \tag{2.30}$$

$$|v| \leq g(x,y)\left|u^p\right| \tag{2.31}$$

where $u(x,y)$ and $g(x,y)$ be nonnegative continuous functions defined on a domain D $0 < p < 1 \cdot p + q = 1, q > 0$. The Equation (2.28) with (2.29) is equivalent to the integral equation

$$u(x,y) = a_1(x) + a_2(y) + \int_{x_0}^{x}\int_{y_0}^{y} A(s,t,u(s,t),v(s,t))\,dtds \tag{2.32}$$

Let $u(x,y)$ be any solution of (2.28) with (2.29) and taking absolute values of both sides, we get

$$|u(x,y)| = |a_1(x) + a_2(y)| + \int_{x_0}^{x}\int_{y_0}^{y} \left|A(s,t,u(s,t),v(s,t))\right|dtds, \tag{2.33}$$

Using (2.30)-(2.32) in (2.33) and assuming that $|a_1(x)|+|a_2(y)|\le w(x,y)$, where $w(x,y)$ be a positive continuous and nondecreasing function defined in the respective domain, we have

$$|u(x,y)|\le w(x,y)+\int_{x_0}^{x}\int_{y_0}^{y}|u(s,t)|\,dt\,ds+\int_{x_0}^{x}\int_{y_0}^{y}g(s,t)|u^p(s,t)|\,dt\,ds$$

Then $|u(x,y)|\le E^*(x,y)\left[w(x,y)+r(x,y)\right]$,

where $E^*(x,y)=\left[1+(1-p)\int_0^x\int_0^y a(s,t)m^{-q}(s,t)\,dt\,ds\right]^{\frac{1}{1-p}}$, and $r(x,y)$ is the maximal solution of

$$r_{xy}(x,y)=E^*(s,t)\left[w(s,t)+r(s,t)\right],\ r_x(x,0)=0,\ r(0,y)=0$$

The remaining proof will be the same as the proof of Theorem 2.1 with suitable modifications.

We note that Theorem 2.1 can be used to study the stability, boundedness and continuous dependence of the solutions of (2.28).

REFERENCES

1. Gronwall, T.H. (1919) Note on the Derivatives with Respect to a Parameter of Solutions of a System of Differential Equations. Annals of Mathematics, 20, 292-296.

2. Guiliano, L. (1946) Generalazzioni di un lemma di Gronwall. Rend.Acad.Lecci, 1264-1271.

3. Bellman, R. (1943) The Stability of Solutions of Linear Differential Equations. Duke Mathematical Journal, 10, 643-647.

4. Nemyckii, V.V and Stepanov, V.V. (1947) Qualitative Theory of Differential Equations (Russian), Moscow, OGIZ.

5. Bihari, I. (1956) A Generalization of a Lemma of Bellman and Its Application to Uniqueness Problem of Differential Equations. Acta Mathematica Hungarica, 7, 71-94.

6. Langenhop, C.E. (1960) Bounds on the Norm of a Solution of a General Differential Equation. Proceedings of the American Mathematical Society, 11, 795-799.

7. Beckenbach, E.F and Bellman, R. (1961) Inequalities. Springer-Verlag, New York.

8. Bainov, D. and Simeonov, P. (1992) Integral Inequalities and Applications. Kluwer Academic Publishers, Dordrecht.

9. Dragomir, S.S and Kim, Y.H. (2002) On Certain New Integral Inequalities and Their Applications. Journal of Inequalities in Pure and Applied Mathematics, 3, 1-8.

10. Dragomir, S.S and Kim, Y.H. (2003) Some Integral Inequalities for Functions of Two Variables. Electronic Journal of Differential Equations, 10, 1-13.

11. Mitrinovic, D.S, Pecaric, J.E. and Fink, A.M. (1991) Inequalities Involving Functions and Their Integrals and Derivatives. Kluwer Academic Publishers, Dordrecht, Boston, London.

12. Pachpatte, B.G. (2001) On Some Fundamental Integral Inequalities and Their Discrete Analogues. Journal of Inequalities in Pure and Applied Mathematics, 2, Article 15.

13. Pachpatte, B.G. (1994b) On Some Fundamental Integral Inequalities Arising in the Theory of Differential Equations. Chinese Journal of Mathematics, 22, 261-273.

14. Pachpatte, B.G. (1996) Comparison Theorems Related to a Certain Inequality Used in the Theory of Differential Equations. Soochow Journal of Mathematics, 22, 383-394.

CHAPTER 3

Two-dimensional Chebyshev hybrid functions and their applications to integral equations

Zahra JafariBehbahani*, **Masood Roodaki**

Department of Mathematics, Marvdasht Branch,
Islamic Azad University, Marvdasht, Iran

ABSTRACT

A combination of bivariate Chebyshev polynomials and two-dimensional block-pulse functions are introduced and applied for approximating the numerical solution of two-dimensional Fredholm integral equations. All calculations in this approach would be easily implemented. The method has the advantage of reducing computational burden. The convergence analysis is given. Some numerical examples are provided to illustrate the accuracy and computational efficiency of the proposed method.

KEYWORDS

Chebyshev polynomials of the first kind; Block-pulse functions; Two-dimensional hybrid Chebyshev and block-pulse functions; Two-dimensional linear Fredholm integral equations; Direct method

1. INTRODUCTION

Two-dimensional Fredholm integral equations appear in mathematical modeling of various phenomena in physics and engineering. By using potential theorem, three dimensional Laplace's equation with boundary condition can be transformed into two dimensional boundary integral equation with weakly singular kernel function (Kress, 1989). The two dimensional images deblurring problem can be modeled as a two dimensional integral equation with smooth kernel function (Chan and Ng, 1996 and Rajan and Chaudhuri, 2003). The transport equation is a linear case of the Boltzmann equation with wide applications in physics and engineering. This equation can be formulated as a two-dimensional Fredholm integral equation (Kadem and Baleanu, 2012). The two-dimensional Fredholm integral equation methods are developed and used for electromagnetic analysis, specifically for antennas and radar scattering (Volakis and Sertel, 2012). In linear case, these equations have the following form

$$f(s,t) = g(s,t) + \int_0^{T_2}\int_0^{T_1} k(s,t,x,y) f(x,y)\,dx\,dy, \quad (x,y) \in D, \tag{1}$$

where $f(s,t)$ is an unknown scalar valued function defined on $D = [0,T_1] \times [0,T_2]$. The functions $k(s,t,x,y)$ and $g(s,t)$ are given functions defined on

$W=\{(s,t,x,y): 0 \le x \le s \le T_1, 0 \le y \le t \le T_2\}$, and D, respectively (Atkinson, 1997 and Delves and Mohammed, 1985). Since any finite interval $[a,b]$ can be transformed to [0,1] by linear maps, it is supposed that $[0,T_1] = [0,T_2] = [0,1]$, without any loss of generality.

In comparison to one-dimensional integral equations, few numerical methods are known for approximating solution of Eq. (1). Hanson and Phillips (1978) proposed numerical solution of two-dimensional integral equations using linear elements. Guoqiang and Jiong (2001) used extrapolation for computing Nyström solution of two-dimensional nonlinear Fredholm integral equations. Gaussian radial basis functions was used for (1) by Alipanah and Esmaeili (2011). Some numerical methods based on piecewise polynomial interpolation was presented (Xie and Lin, 2009 and Liang and Lin, 2010).Maleknejad and Mahdiani (2011) used two-dimensional block-pulse functions for solving mixed Volterra-Fredholm integral equations and nonlinear two-dimensional integral equations of the first kind (Maleknejad et al., 2010). Recently, the direct approaches for estimating numerical solution of two-dimensional Volterra-Fredholm and mixed integral equations were proposed using triangular orthogonal functions (Babolian et al., 2010 and Maleknejad and JafariBehbahani, 2012).

In this paper, a new set of basis functions are constructed by combination of bivariate Chebyshev polynomials and two-dimensional block-pulse functions. A proposed name for these functions may be two-dimensional hybrid Chebyshev and block-pulse functions (2D-ChBPFs). Then solutions of two-dimensional linear Fredholm integral equations are computed approximately in a direct approach.

The proposed method, reduces a two-dimensional linear Fredholm integral equation to a system of algebraic equations. High accuracy of Chebyshev polynomials in expansion of functions leads to a good approximate solution for (1). Moreover, essential properties of block-pulse functions is due to setting up the algebraic system in a simple manner. The uniform convergence analysis and accuracy estimation of the method is presented. Finally, we check the proposed method on some examples to show its accuracy and superiority.

2. PRELIMINARIES

The well-known Chebyshev polynomials of the first kind are defined by

$$T_n(t) = \cos\left(n\cos^{-1}t\right), \quad n \geq 0.$$

They have several applications in approximation theory and numerical analysis (Chihara, 1978). A set of Chebyshev polynomials is an orthogonal set with respect to the weight function $w(t) = \frac{1}{\sqrt{1-t^2}}$ on the interval [−1,1], that is

$$< T_i(t), T_j(t) > = \int_{-1}^{1} \frac{T_i(t)T_j(t)}{\sqrt{1-t^2}}dt = \begin{cases} \pi, & i=j=0, \\ \frac{\pi}{2}, & i=j>0, \\ 0, & i \neq j. \end{cases}$$

They may be derived from the following recursive formula

$$\begin{aligned} &T_0(t) = 1, \\ &T_1(t) = t, \\ &T_{m+1}(t) = 2tT_m(t) - T_{m-1}(t), \quad m = 1,2,3,\cdots. \end{aligned}$$

In a two-dimensional case, we can consider the bivariate Chebyshev polynomials on region [−1,1] × [−1,1] as

$$T_{i,j}(s,t) = T_i(s)T_j(t), \quad i,j \geq 0, \tag{2}$$

and any function $f(s,t)$ defined on $[-1,1] \times [-1,1]$ can be expanded with respect to bivariate Chebyshev polynomials as

$$f(s,t) = \sum_{j=0}^{\infty}\sum_{i=0}^{\infty} a_{ij} T_{i,j}(s,t) \cong \sum_{j=0}^{M_2}\sum_{i=0}^{M_1} a_{ij} T_{i,j}(s,t),$$

for arbitrary positive integers M_1 and M_2 (Hadizadeh and Asgari, 2005). The coefficients *aij* may be computed as

$$a_{i,j} = \frac{< T_i(s), < f(s,t), T_j(t) > >}{< T_i(s), T_i(s) > < T_j(t), T_j(t) >}, \quad i,j = 0,1,2,\cdots,$$

where <·,·> denotes the inner product with respect to suitable weight function $w(t) = \frac{1}{\sqrt{1-t^2}}$ or $w(s) = \frac{1}{\sqrt{1-s^2}}$.

On the other hand, in an $(N_1 \times N_2)$-set of two dimensional block-pulse functions (2D-BPFs) on the region $[0,1) \times [0,1)$, each component is defined as

$$\phi_{n_1,n_2}(s,t) = \begin{cases} 1, & \frac{n_1-1}{N_1} \le s < \frac{n_1}{N_1}, \frac{n_2-1}{N_2} \le t < \frac{n_2}{N_2}, \\ 0, & otherwise, \end{cases}$$

where $n_1 = 1,2,\cdots,N_1$ and $n_2 = 1,2,\cdots,N_2$, for arbitrary positive integers N_1 and N_2 (Ganti, 1983). Some important properties of 2D-BPFs such as disjointness, orthogonality, and completeness are described in (Maleknejad et al., 2010 and Ganti, 1983).

3. HYBRID CHEBYSHEV POLYNOMIALS AND 2D-BPFS

In recent years, one-dimensional hybrid functions have been investigated in several studies as basis functions to estimate solutions of various equations.

Now, we can combine the bivariate Chebyshev polynomials and two-dimensional block-pulse functions to construct a new set of orthogonal functions over the region [0,1) × [0,1) as

$$b_{n_1,m_1;n_2,m_2}(s,t) = \begin{cases} T_{m_1,m_2}(2N_1 s - 2n_1 + 1, 2N_2 t - 2n_2 + 1), & \dfrac{n_1 - 1}{N_1} \le s < \dfrac{n_1}{N_1}, \\ & \dfrac{n_2 - 1}{N_2} \le t < \dfrac{n_2}{N_2}, \\ 0, & otherwise, \end{cases} \tag{3}$$

in which

N_1 denotes number of dissections of interval [0,1) over s axis,

M_1 denotes number of Chebyshev polynomials considered for variable s,

N_2 denotes number of dissections of interval [0,1) over t axis,

M_2 denotes number of Chebyshev polynomials considered for variable t,

and

$$n_1 = 1, 2, \cdots, N_1,$$
$$m_1 = 0, 1, \cdots, M_1 - 1,$$
$$n_2 = 1, 2, \cdots, N_2,$$
$$m_2 = 0, 1, \cdots, M_2 - 1.$$

Hence the region $[0,1) \times [0,1)$ is divided to $(N_1 \times N_2)$-subregions and $M_1 \times M_2$ bivariate Chebyshev polynomials are constructed on each of them. Throughout this paper, we assume that $n_1, m_1, n_2, m_2, N_1, M_1, N_2$ and M_2 are the same as those defined above. It is natural that we call the polynomials in (3), two-dimensional hybrid of Chebyshev and block-pulse functions (2D-ChBPFs). Some properties of 2D-ChBPFs, which we need in future sections, are as follows.

i)The 2D-ChBPFs may be written as

equation(4)

$h_{n1,m1;n2,m2}(s,t)=bs_{n1,m1}(s)\cdot bt_{n2,m2}(t)$,

where

$$bs_{n_1,m_1}(s)=\begin{cases} T_{m_1}(2N_1s-2n_1+1), & \frac{n_1-1}{N_1}\leq s<\frac{n_1}{N_1}, \\ 0, & otherwise, \end{cases}$$

and

$$bt_{n_2,m_2}(t)=\begin{cases} T_{m_2}(2N_2t-2n_2+1), & \frac{n_2-1}{N_2}\leq t<\frac{n_2}{N_2}, \\ 0, & otherwise, \end{cases}$$

are the one-dimensional hybrid Chebyshev block-pulse functions over variables s and t with respect to weight functions $ws_{n_1}(s)=\frac{1}{\sqrt{1-(2N_1s-2n_1+1)^2}}$ and $wt_{n_2}(t)=\frac{1}{\sqrt{1-(2N_2t-2n_2+1)^2}}$, respectively.

ii)It is obvious that 2D-ChBPFs defined on different subregions are disjoint; that is, for possible choice of indices we have

$$b_{p_1,i_1;q_1,j_1}(s,t)b_{p_2,i_2;q_2,j_2}(s,t)=0, \quad p_1 \neq p_2 \ or \ q_1 \neq q_2. \tag{5}$$

iii)The 2D-ChBPFs defined on (n_1,n_2)-th subregion of [0,1) × [0,1) are orthogonal with respect to weight function

$$W_{n_1,n_2}(s,t)=\frac{1}{\sqrt{1-(2N_1s-2n_1+1)^2}\sqrt{1-(2N_2t-2n_2+1)^2}}. \tag{6}$$

It means that

$$\int_{-1}^{1}\int_{-1}^{1} b_{n_1,i_1;n_2,j_1}(s,t)b_{n_1,i_2;n_2,j_2}(s,t)\omega_{n_1,n_2}(s,t)dsdt=\lambda_{i_1}\delta_{i_1,i_2}\cdot\lambda_{j_1}\delta_{j_1,j_2},$$

where

$$\lambda_k=\begin{cases}\pi, & k=0,\\ \frac{\pi}{2}, & k>0,\end{cases}$$

and δ denotes the Kronecker delta function.

iv)The 2D-ChBPFs defined on (n_1,n_2)-th subregion may be simplified in the followingM_1M_2-vector

$$B_{n_1,n_2}(s,t)=\begin{bmatrix} b_{n_1,0;n_2,0}, & b_{n_1,0;n_2,1}, & \cdots, & b_{n_1,0;n_2,M_2-1} \\ b_{n_1,1;n_2,0}, & b_{n_1,1;n_2,1}, & \cdots, & b_{n_1,1;n_2,M_2-1} \\ \vdots & \vdots & & \vdots \\ b_{n_1,M_1-1;n_2,0}, & b_{n_1,M_1-1;n_2,1}, & \cdots, & b_{n_1,M_1-1;n_2,M_2-1}\end{bmatrix}^T. \tag{7}$$

So

$$\mathbf{B}(s,t)=\begin{bmatrix} B_{1,1}^T, & B_{1,2}^T, & \cdots & B_{1,N_2}^T \\ B_{2,1}^T, & B_{2,2}^T, & \cdots & B_{2,N_2}^T \\ \vdots & \vdots & & \vdots \\ B_{N_1,1}^T, & B_{N_1,2}^T, & \cdots & B_{N_1,N_2}^T \end{bmatrix}^T, \quad (8)$$

is an $N_1N_2M_1M_2$-vector concluding all of 2D-ChBPFs. In Eqs. (7) and (8) the term (s,t) is canceled for convenience. It is simple to verify that the function $b_{n1,m1;n2,m2}(s,t)$ is attached in *k*-th component of vector B where

$$k=(n_1-1)M_1M_2N_2+(n_2-1)M_1M_2+m_1M_2+m_2+1. \quad (9)$$

4. FUNCTION EXPANSION

A function *f*(*s,t*) defined on [0,1) × [0,1) may be expanded with respect to 2D-ChBPFs as

$$\begin{aligned} f(s,t) &\cong \sum_{n_1=1}^{N_1}\sum_{n_2=1}^{N_2}\sum_{m_1=0}^{M_1-1}\sum_{m_2=0}^{M_2-1} d_{n_1,m_1,n_2,m_2} b_{n_1,m_1;n_2,m_2}(s,t), \\ &= F^T \cdot \mathbf{B}(s,t), \end{aligned}$$

where B(*s,t*) is defined in (8) and the $N_1N_2M_1M_2$-vector F contains the coefficients $d_{n1,m1,n2,m2}$ with the same indices in B. Under notation (9), orthogonality of 2D-ChBPFs implies that

$$F_k = d_{n_1,m_1;n_2,m_2} = \frac{< Bs_{n_1,m_1}(s), < f(s,t), Bt_{n_2,m_2}(t) > >}{< Bs_{n_1,m_1}(s), Bs_{n_1,m_1}(s) > \cdot < Bt_{n_2,m_2}, Bt_{n_2,m_2} >}, \quad (10)$$

in which <·,·> denotes the inner product with respect to suitable weight functions $\omega s_{n1}(s)$ or $\omega t_{n2}(t)$. Since

$$< Bs_{n_1,m_1}(s), Bs_{n_1,m_1}(s) > = \int_{\frac{n_1-1}{N_1}}^{\frac{n_1}{N_1}} \frac{Bs_{n_1,m_1}(s)Bs_{n_1,m_1}(s)}{\sqrt{1-(2N_1s-2n_1+1)^2}}ds = \frac{\lambda_{m_1}}{2N_1},$$

$$< Bt_{n_2,m_2}(t), Bt_{n_2,m_2}(t) > = \int_{\frac{n_2-1}{N_2}}^{\frac{n_2}{N_2}} \frac{Bt_{n_2,m_2}(t)Bt_{n_2,m_2}(t)}{\sqrt{1-(2N_2t-2n_2+1)^2}}dt = \frac{\lambda_{m_2}}{2N_2},$$

we have

$$F_k = \frac{4N_1N_2}{\lambda_{m_1}\lambda_{m_2}} \cdot \int_{\frac{n_1-1}{N_1}}^{\frac{n_1}{N_1}} \left(\frac{Bs_{n_1,m_1}(s)}{\sqrt{1-(2N_1s-2n_1+1)^2}} \cdot \int_{\frac{n_2-1}{N_2}}^{\frac{n_2}{N_2}} \frac{f(s,t)Bt_{n_2,m_2}(t)}{\sqrt{1-(2N_2t-2n_2+1)^2}}dt \right) ds,$$

for $k = 1,2,\cdots,N_1N_2M_1M_2$.

Furthermore, the expansion of function $k(s,t,x,y)$ may be written as

$$k(s,t,x,y) = B^T(s,t) \cdot K \cdot B(x,y), \tag{11}$$

where K is an $(N_1N_2M_1M_2 \times N_1N_2M_1M_2)$-matrix with

$$K_{ij} = \left(\frac{4N_1N_2}{\lambda_{m_1}\lambda_{m_2}}\right)^2 \cdot \int_{\frac{n_1-1}{N_1}}^{\frac{n_1}{N_1}} \left(\frac{Bs_{n_1,m_1}(s)}{\sqrt{1-(2N_1s-2n_1+1)^2}} \right.$$

$$\cdot \int_{\frac{n_2-1}{N_2}}^{\frac{n_2}{N_2}} \left(\frac{Bt_{n_2,m_2}(t)}{\sqrt{1-(2N_2t-2n_2+1)^2}} \right.$$

$$\cdot \int_{\frac{p_1-1}{N_1}}^{\frac{p_1}{N_1}} \left(\frac{Bs_{p_1,q_1}(x)}{\sqrt{1-(2N_1x-2p_1+1)^2}} \right.$$

$$\left. \left. \left. \cdot \int_{\frac{p_2-1}{N_2}}^{\frac{p_2}{N_2}} \frac{k(s,t,x,y)Bt_{p_2,q_2}(y)}{\sqrt{1-(2N_2y-2p_2+1)^2}} dy \right) dx \right) dt \right) ds,$$

in which

$$i = (n_1-1)M_1M_2N_2 + (n_2-1)M_1M_2 + m_1M_2 + m_2 + 1,$$
$$j = (p_1-1)M_1M_2N_2 + (p_2-1)M_1M_2 + q_1M_2 + q_2 + 1,$$

for $n_1,p_1 = 1,2,\cdots,N_1$, $n_2,p_2 = 1,2,\cdots,N_2$, $m_1,q_1 = 0,1,\cdots,M_1-1$ and $m_2,q_2 = 0,1,\cdots,M_2-1$.

5. ERROR ANALYSIS

In this section, we present the uniform convergence and accuracy estimation of the presented method.

THEOREM 1.

Suppose that for the function f:[0,1] × [0,1]→R, *the partial derivatives* $\partial^{N_1+1}f(s,t)/\partial s^{N_1+1}$ *and* $\partial^{N_2+1}f(s,t)/\partial t^{N_2+1}$ *exist and are continuous for all* $(s,t)\in[0,1]\times[0,1]$. *Then* $F^T\cdot B(s,t)$ *approximate* $f(s,t)$ *with the following mean error bound*

$$\left\|f(s,t)-F^T\cdot\mathbf{B}(s,t)\right\| \le \frac{1}{2^{N_1}(N_1+1)!} max_{\xi\in[0,1]} \frac{\partial^{N_1+1}f(\xi,t)}{\partial\xi^{N_1+1}} + \frac{\delta}{2^{N_2}(N_2+1)!} max_{(\xi,\eta)\in[0,1]\times[0,1]} \frac{\partial^{N_2+1}f(\xi,\eta)}{\partial\eta^{N_2+1}}.$$

PROOF.

Using 2D-ChBPFs, the region [0,1] × [0,1] is divided into $(N_1 \times N_2)$-subregions $I_{n_1,n_2} = [\frac{n_1-1}{N_1},\frac{n_1}{N_1}]\times[\frac{n_2-1}{N_2},\frac{n_2}{N_2}]$ and the function $f(s,t)$ is approximated on each ones. Since $F^T B(s,t)$ is a polynomial which approximate $f(s,t)$ with minimum mean error, we have

$$\left\|f(s,t)-F^T\cdot\mathbf{B}(s,t)\right\|^2 = \int_0^1\int_0^1 W(s,t)\left(f(s,t)-F^T\mathbf{B}(s,t)\right)^2 dtds$$

$$= \sum_{n_1=1}^{N_1}\sum_{n_2=1}^{N_2}\int_{\frac{n_1-1}{N_1}}^{\frac{n_1}{N_1}}\int_{\frac{n_2-1}{N_2}}^{\frac{n_2}{N_2}} W_{n_1,n_2}(s,t)\left(f(s,t)-F^T\mathbf{B}(s,t)\right)^2 dtds$$

$$\le \sum_{n_1=1}^{N_1}\sum_{n_2=1}^{N_2}\int_{\frac{n_1-1}{N_1}}^{\frac{n_1}{N_1}}\int_{\frac{n_2-1}{N_2}}^{\frac{n_2}{N_2}} W_{n_1,n_2}(s,t)(f(s,t)-p(s,t))^2 dtds,$$

where $p(s,t)$ is any interpolating polynomial of $f(s,t)$ on $I_{n1,n2}$ at Chebyshev's zeroes. In general, we have

$$\left|f(s,t)-p(s,t)\right| \le \frac{|\omega_{N_1}(s)|}{(N_1+1)!} max_{\xi\in[0,1]}\left|\frac{\partial^{N_1+1}f(\xi,t)}{\partial\xi^{N_1+1}}\right| + \delta\frac{|\omega_{N_2}(t)|}{(N_2+1)!} max_{(\xi,\eta)\in[0,1]\times[0,1]}\frac{\partial^{N_2+1}f(\xi,\eta)}{\partial\eta^{N_2+1}},$$

where $\omega l(\tau) = (\tau - \tau_0)\cdots(\tau - \tau l)$,
and $\delta = \max_{s\in[0,1]} \Lambda_{N_1}(s) = \sum_{i=0}^{N_1}|l_{i,N_1}(s)|$, $l_{i,N1}(s)$ are Lagrange's polynomial with degree of N_1(Atkinson, 1997). So this error bound for $p(s,t)$ at Chebyshev's zeroes gives

$$\left\|f(s,t)-F^T\cdot\mathbf{B}(s,t)\right\|^2 \le \sum_{n_1=1}^{N_1}\sum_{n_2=1}^{N_2}\int_{\frac{n_1-1}{N_1}}^{\frac{n_1}{N_1}}\int_{\frac{n_2-1}{N_2}}^{\frac{n_2}{N_2}} W_{n_1,n_2}(s,t)$$

$$\cdot\left(\frac{1}{2^{N_1}(N_1+1)!} max_{\xi_{n_1}\in\left[\frac{n_1-1}{N_1},\frac{n_1}{N_1}\right]}\left|\frac{\partial^{N_1+1}f(\xi_{n_1},t)}{\partial\xi_{n_1}^{N_1+1}}\right|\right.$$

$$\left.+\delta\frac{1}{2^{N_2}(N_2+1)!} max_{(\xi_{n_1},\eta_{n_2})\in I_{n_1n_2}}\left|\frac{\partial^{N_2+1}f(\xi_{n_1},\eta_{n_2})}{\partial\eta_{n_2}^{N_2+1}}\right|\right)^2 dtds$$

$$\le \sum_{n_1=1}^{N_1}\sum_{n_2=1}^{N_2}\int_{\frac{n_1-1}{N_1}}^{\frac{n_1}{N_1}}\int_{\frac{n_2-1}{N_2}}^{\frac{n_2}{N_2}} W_{n_1,n_2}(s,t)$$

$$\cdot\left(\frac{1}{2^{N_1}(N_1+1)!} max_{\xi\in[0,1]}\left|\frac{\partial^{N_1+1}f(\xi,t)}{\partial\xi^{N_1+1}}\right|\right.$$

$$\left.+\delta\frac{1}{2^{N_2}(N_2+1)!} max_{(\xi,\eta)\in[0,1]\times[0,1]}\left|\frac{\partial^{N_2+1}f(\xi,\eta)}{\partial\eta^{N_2+1}}\right|\right)^2 dtds$$

$$=\int_0^1\int_0^1 W(s,t)\left(\frac{1}{2^{N_1}(N_1+1)!} max_{\xi\in[0,1]}\left|\frac{\partial^{N_1+1}f(\xi,t)}{\partial\xi^{N_1+1}}\right|\right.$$

$$\left.+\frac{\delta}{2^{N_2}(N_2+1)!} max_{(\xi,\eta)\in[0,1]\times[0,1]}\left|\frac{\partial^{N_2+1}f(\xi,\eta)}{\partial\eta^{N_2+1}}\right|\right)^2 dtds$$

$$=\mathrm{P}\frac{1}{2^{N_1}(N_1+1)!} max_{\xi\in[0,1]}\frac{\partial^{N_1+1}f(\xi,t)}{\partial\xi^{N_1+1}}$$

$$+\frac{\delta}{2^{N_2}(N_2+1)!} max_{(\xi,\eta)\in[0,1]\times[0,1]}\frac{\partial^{N_2+1}f(\xi,\eta)}{\partial\eta^{N_2+1}}\mathrm{P}^2.$$

6. OPERATIONAL MATRIX FOR INTEGRATION OF PRODUCT VECTORS

It is immediately concluded from Eqs. (5), (7) and (8) that

$$\mathbf{B}(s,t)\cdot\mathbf{B}^T(s,t)=\begin{bmatrix} B_{1,1}(s,t)\\ B_{1,2}(s,t)\\ \vdots\\ B_{N_1,N_2}(s,t)\end{bmatrix}\cdot\left[B_{1,1}^T(s,t),\ B_{1,2}^T(s,t),\ \ldots,\ B_{N_1,N_2}^T(s,t)\right]$$

$$=\begin{bmatrix} B_{1,1}(s,t)\cdot B_{1,1}^T(s,t) & 0 & \cdots & 0\\ 0 & B_{2,2}(s,t)\cdot B_{2,2}^T(s,t) & \ddots & \vdots\\ \vdots & \ddots & \ddots & 0\\ 0 & \cdots & 0 & B_{N_1,N_2}(s,t)\cdot B_{N_1,N_2}^T(s,t)\end{bmatrix}.$$

So,

$$\int_0^1\int_0^1 \mathbf{B}(s,t)\cdot\mathbf{B}^T(s,t)\,dtds\cong D, \tag{12}$$

where D is an $(N_1N_2M_1M_2 \times N_1N_2M_1M_2)$-diagonal matrix of the form

$$D=\begin{bmatrix} D_1 & 0 & \cdots & 0\\ 0 & D_2 & \ddots & \vdots\\ \vdots & \ddots & \ddots & 0\\ 0 & \cdots & 0 & D_{N_1N_2}\end{bmatrix}. \tag{13}$$

The components of each diagonal block matrices of D can be computed as follows

$$(D_\eta)_{\rho,\rho}=\frac{1}{4N_1N_2}\int_{-1}^{1}\int_{-1}^{1} T_{m_1}^2(s)\,T_{m_2}^2(t)\,dtds,$$

in which $\eta = n_1n_2$, $\rho = m_1M_2 + (m_2 + 1)$, and m_1, m_2, n_1 and n_2 are suitable indices.

Moreover, let Ds_{n1} be the following $M_1 \times M_1$ matrix

$$(Ds_{n_1})_{m_1,m_1} = \frac{1}{2N_1}\int_{-1}^{1} T_{m_1}^2(s)ds, 2cm \begin{matrix} m_1 = 0,1,\cdots,M_1-1, \\ n_1 = 1,2,\cdots,N_1, \end{matrix} \tag{14}$$

and similarly the $M_2 \times M_2$ matrix Dt_{n2} be

$$(Dt_{n_2})_{m_2,m_2} = \frac{1}{2N_2}\int_{-1}^{1} T_{m_2}^2(t)dt, 2cm \begin{matrix} m_2 = 0,1,\cdots,M_2-1, \\ n_2 = 1,2,\cdots,N_2, \end{matrix} \tag{15}$$

These two matrices can be symbolically written as

$$Ds = \begin{bmatrix} Ds_1 & 0 & \cdots & 0 \\ 0 & Ds_2 & \ddots & \vdots \\ \vdots & \ddots & \ddots & 0 \\ 0 & \cdots & 0 & Ds_{N_1} \end{bmatrix} \quad and$$

$$Dt = \begin{bmatrix} Dt_1 & 0 & \cdots & 0 \\ 0 & Dt_2 & \ddots & \vdots \\ \vdots & \ddots & \ddots & 0 \\ 0 & \cdots & 0 & Dt_{N_2} \end{bmatrix},$$

respectively (Tavassoli Kajani and Hadi Vencheh, 2005). Now, the matrix D in Eq. (13)can be computed as

$$D = Ds\overline{\otimes}Dt,$$

where $\overline{\otimes}$ denotes the Kronecker block product defined for two arbitrary block matrices$P$ and Q by multiplying each block of P in all blocks of Q

$$P\overline{\otimes}Q = \begin{bmatrix} P_1 & P_2 \\ P_3 & P_4 \end{bmatrix}\overline{\otimes}\begin{bmatrix} Q_1 & Q_2 \\ Q_3 & Q_4 \end{bmatrix} = \begin{bmatrix} P_1\otimes Q_1 & P_1\otimes Q_2 & P_2\otimes Q_1 & P_2\otimes Q_2 \\ P_1\otimes Q_3 & P_1\otimes Q_4 & P_2\otimes Q_3 & P_2\otimes Q_4 \\ P_3\otimes Q_1 & P_3\otimes Q_2 & P_4\otimes Q_1 & P_4\otimes Q_2 \\ P_3\otimes Q_3 & P_3\otimes Q_4 & P_4\otimes Q_3 & P_4\otimes Q_4 \end{bmatrix}, \tag{16}$$

and $\otimes$ denotes the Kronecker product defined for two arbitrary matrices A and B as

$A \otimes B = (a_{ij}B)$. (17)

7. SOLVING 2D-FREDHOLM INTEGRAL EQUATIONS

In this section, we present an effective method for solving two-dimensional Fredholm integral equation (2D-FIE) using 2D-ChBPFs.

Consider the following 2D-FIE

$$f(s,t) = g(s,t) + \int_0^1 \int_0^1 k(s,t,x,y) f(x,y)\,dx\,dy, \tag{18}$$

by the method mentioned in previous section, we can approximate functions $f(s,t)$,$g(s,t)$ and $k(s,t,x,y)$ as follows

$$\begin{aligned} f(s,t) &\cong F^T \cdot \mathbf{B}(s,t) = \mathbf{B}^T(s,t) \cdot F, \\ g(s,t) &\cong G^T \cdot \mathbf{B}(s,t) = \mathbf{B}^T(s,t) \cdot G, \\ k(s,t,x,y) &\cong \mathbf{B}^T(s,t) \cdot K \cdot \mathbf{B}(x,y), \end{aligned} \tag{19}$$

where B(s,t) is defined in (8), and $N_1N_2M_1M_2$-vectors F and G are 2D-ChBPFs coefficients of $f(s,t)$ and $g(s,t)$, respectively. The $(N_1N_2M_1M_2 \times N_1N_2M_1M_2)$-matrix K is 2D-ChBPFs coefficients of $k(s,t,x,y)$.

To approximate the integral part in (18), from (19) and (12), we have

$$\begin{aligned} \int_0^1 \int_0^1 k(s,t,x,y) f(x,y)\,dx\,dy &\cong \int_0^1 \int_0^1 \mathbf{B}^T(s,t) \cdot K \cdot \mathbf{B}(x,y) \cdot \mathbf{B}^T(x,y) \cdot F\,dx\,dy \\ &= \mathbf{B}^T(s,t) \cdot K \cdot \int_0^1 \int_0^1 \mathbf{B}(x,y) \cdot \mathbf{B}^T(x,y)\,dx\,dy \cdot F \\ &= \mathbf{B}^T(s,t) \cdot K \cdot D \cdot F, \end{aligned} \tag{20}$$

with D defined in (13).

Substituting (19) and (20) in (18) and replacing ≅ with =, follow

$$BT(s,t)\cdot F=BT(s,t)\cdot G+BT(s,t)\cdot K\cdot D\cdot F,$$

or

equation(21)

$$(I-K\cdot D)F=G. \quad (21)$$

Eq. (21) is a linear system of $N_1N_2M_1M_2$ algebraic equations. $N_1N_2M_1M_2$ components of F are unknown, and can be obtained by solving this system. Hence, an approximate solution

$$f(s,t)\cong FT\cdot B(s,t), \quad (22)$$

can be computed for Eq. (18).

8. NUMERICAL EXAMPLES

The method presented in this paper is applied to some examples. In first example, the construction of matrices K and D, and vector G is illustrated. In second example, the error in each subregion is computed for various N_1,N_2 by

$$e_{n_1,n_2} = \left\| \bar{f}_{n_1,n_2}(s,t) - f(s,t) \right\|_2 = \left[\int_{\frac{n_1-1}{N_1}}^{\frac{n_1}{N_1}} \int_{\frac{n_2-1}{N_2}}^{\frac{n_2}{N_2}} \left(\bar{f}_{n_1,n_2}(s,t) - f(s,t) \right)^2 dt ds \right]^{\frac{1}{2}},$$

(23)

where $f(s,t)$ and $\bar{f}_{n_1,n_2}(s,t)$ are the exact and approximate solutions of the integral equation, respectively. So the total error on $[0,1] \times [0,1]$ may be obtained by

$$e_{total}(s,t) = \left(\sum_{n_1=1}^{N_1} \sum_{n_2=1}^{N_2} e_{n_1,n_2}(s,t)^2 \right)^{\frac{1}{2}}. \tag{24}$$

In the third example, our results are compared with the exact solutions by calculating the following error function

$$e_{grids}(s,t) = \left| f(s,t) - \bar{f}_{n_1,n_2}(s,t) \right|, \quad \begin{array}{l} \frac{n_1-1}{N_1} \le s < \frac{n_1}{N_1}, \\ \frac{n_2-1}{N_2} \le t < \frac{n_2}{N_2}. \end{array}$$

The values of *egrids*(*s*,*t*) over the set

$D_{grids} = \{(0.0,0.0),(0.1,0.1),(0.2,0.2),\cdots,(1.0,1.0)\},$

are computed for different values of M_1, M_2, N_1 and N_2, and collected in Table 2.

Table 1. The numerical results for example 2.

M_1,M_2	N_1,N_2		e_{n_1,n_2}				e_{total}
3,3	2,2		$n_2 = 1$	$n_2 = 2$			
		$n_1 = 1$	0.028996	0.048706			
		$n_1 = 2$	0.048706	0.078818			0.10781
3,3	3,3		$n_2 = 1$	$n_2 = 2$	$n_2 = 3$		
		$n_1 = 1$	0.007184	0.010026	0.013993		
		$n_1 = 2$	0.010026	0.013993	0.019528		
		$n_1 = 3$	0.013993	0.019528	0.027254		0.048431
3,3	4,4		$n_2 = 1$	$n_2 = 2$	$n_2 = 3$	$n_2 = 4$	
		$n_1 = 1$	0.002776	0.003565	0.004577	0.005877	
		$n_1 = 2$	0.003565	0.004577	0.005877	0.007547	
		$n_1 = 3$	0.004577	0.005877	0.007547	0.009690	
		$n_1 = 4$	0.005877	0.007547	0.009690	0.012442	0.027343

Table 2. The numerical results for example 3.

	$e_{grids}(s,t)$		
(s,t)	Case (I) $M_1 = M_2 = 3$ $N_1 = N_2 = 3$	Case (II) $M_1 = M_2 = 3$ $N_1 = N_2 = 4$	Case (III) $M_1 = M_2 = 4$ $N_1 = N_2 = 3$
(0.0,0.0)	0.031497	0.180050	0.204282
(0.1,0.1)	0.007197	0.002258	0.035219
(0.2,0.2)	0.001473	0.005499	0.234782
(0.3,0.3)	0.013315	0.004242	0.178137
(0.4,0.4)	0.004517	0.001486	0.215088
(0.5,0.5)	0.004902	0.004843	0.067356
(0.6,0.6)	0.003704	0.003129	0.343651
(0.7,0.7)	0.004753	0.003830	0.147908
(0.8,0.8)	0.008120	0.004723	0.268182
(0.9,0.9)	0.018734	0.005821	0.285071
(1.0,1.0)	0.018734	0.011023	0.470217

Moreover, the values of *egrids*(s,t) over the region $[0,1] \times [0,1]$ for examples 2 and 3 are displayed respectively in Fig. 1 and Fig. 2, with $M_1 = M_2 = 3$ and $N_1 = N_2 = 4$.

The elements of matrix K and vector G in (21), that is defined by an integral, may be computed numerically by any quadratures. In examples 2 and 3, we applied three-point Gauss-Chebyshev formula for these purpose.

The computations associated with the examples were performed using Matlab 7.0 software on a personal computer.

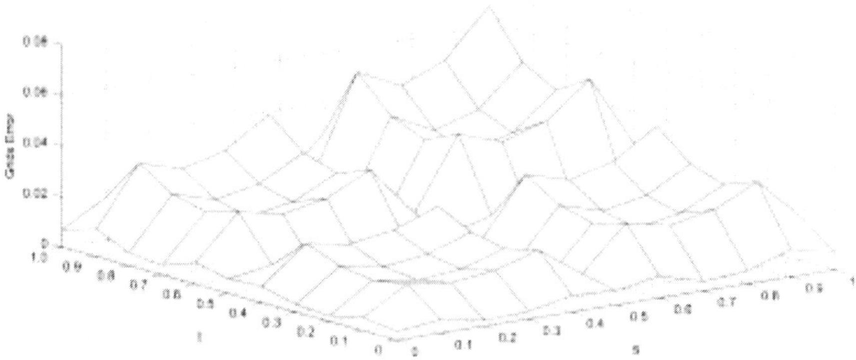

Figure 1. The error function graph for example 2.

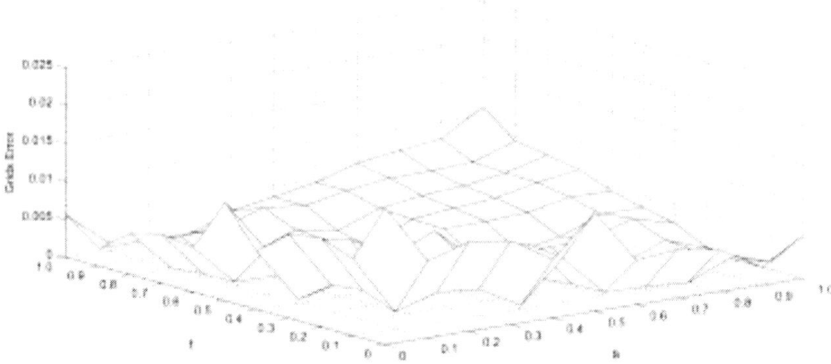

Figure 2. The error function graph for example 3.

EXAMPLE 1.

Consider the following linear Fredholm integral equation

$$f(s,t) = g(s,t) + \int_0^1 \int_0^1 8(st - 2xy) f(x,y)\,dy\,dx, \quad 0 \le s,t \le 1, \tag{25}$$

where

$$g(s,t)=\begin{cases}-6st+2t-3s+\frac{62}{9}, & 0\le s<\frac{1}{2}, 0\le t\le 1,\\ -4st+t+5s+\frac{26}{9}, & \frac{1}{2}\le s\le 1, 0\le t\le 1,\\ 0, & otherwise,\end{cases}$$

with the exact solution

$$f(s,t)=\begin{cases}st+2t-3s, & 0\le s<\frac{1}{2}, 0\le t\le 1,\\ 3st+t+5s-4, & \frac{1}{2}\le s\le 1, 0\le t\le 1,\\ 0, & otherwise.\end{cases}$$

Choosing $N_1 = 2$, $N_2 = 1$ and $M_1 = M_2 = 2$, we have

$$K=\begin{bmatrix}-1 & -2 & -2 & -2 & -5 & -6 & -2 & -2\\ 1 & 0 & 0 & 0 & 1 & 0 & 0 & 0\\ 1 & 0 & 0 & 0 & 1 & 0 & 0 & 0\\ 1 & 0 & 0 & 0 & 1 & 0 & 0 & 0\\ 1 & -2 & -2 & -2 & -3 & -6 & -2 & -2\\ 3 & 0 & 0 & 0 & 3 & 0 & 0 & 0\\ 1 & 0 & 0 & 0 & 1 & 0 & 0 & 0\\ 1 & 0 & 0 & 0 & 1 & 0 & 0 & 0\end{bmatrix},$$

$$D=\begin{bmatrix}\frac{1}{2} & 0 & 0 & 0 & 0 & 0 & 0 & 0\\ 0 & \frac{1}{6} & 0 & 0 & 0 & 0 & 0 & 0\\ 0 & 0 & \frac{1}{6} & 0 & 0 & 0 & 0 & \\ 0 & 0 & 0 & \frac{1}{18} & 0 & 0 & 0 & 0\\ 0 & 0 & 0 & 0 & \frac{1}{2} & 0 & 0 & 0\\ 0 & 0 & 0 & 0 & 0 & \frac{1}{6} & 0 & 0\\ 0 & 0 & 0 & 0 & 0 & 0 & \frac{1}{6} & 0\\ 0 & 0 & 0 & 0 & 0 & 0 & 0 & \frac{1}{18}\end{bmatrix},$$

and

$$G=\left[\frac{115}{18}\ \ \frac{1}{4}\ \ \frac{-3}{2}\ \ \frac{-3}{4}\ \ \frac{203}{36}\ \ 1\ \ \frac{3}{4}\ \ \frac{-1}{2}\right]^{T}.$$

Solving linear system (21), the unknown vector F obtained as:

$$F=\frac{1}{8}[3\ \ 9\ \ -5\ \ 1\ \ 11\ \ 13\ \ 13\ \ 3]^{T},$$

which is expansion coefficients of the exact solution of Eq. (25).

EXAMPLE 2.

For the following Fredholm integral equation

$$f(s,t)=g(s,t)+4\int_{0}^{1}\int_{0}^{1}e^{(s+t)(x+y)}f(x,y)dydx,\quad 0\leq s,t\leq 1,$$

where $g(s,t) = 2e^{s+t+2}-e^{s+t+4}$, with the exact solution $f(s,t) = e^{s+t}$, the values of $e_{total}(s,t)$ and $e_{n1,n2}(s,t)$ for $M_1 = M_2 = 3$ and various N_1 and N_2 are shown in Table 1. In Fig. 1, the graph of *egrids* over the region $[0,1] \times [0,1]$ is illustrated.

EXAMPLE 3.

Consider the following Fredholm integral equation,

$$f(s,t)=g(s,t)-\int_{0}^{1}\int_{0}^{1}\frac{8}{3}cos(s+t+x+y)f(x,y)dydx,$$

where $(s,t) \in [0,1] \times [0,1]$ and $g(s,t) = \frac{1}{3}cos(s+t+4) - \frac{2}{3}cos(s+t+2)$, with the exact solution $f(s,t) = cos(s+t)$. The values of $egrids(s,t)$ over the set $Dgrids$ are shown in Table 2 for three cases.

9. COMMENTS ON THE RESULTS

In this approach, the hybrid bivariate Chebyshev polynomials and two-dimensional block-pulse functions are used to reduce a 2D Fredholm integral equation to a system of algebraic equations without applying any projection method.

Its accuracy and applicability were checked on three examples. Since the integral equation in example 1 was selected by polynomials kernel and known terms, the accurate solution was obtained by suitable choice of M_1 and M_2. Thus, the suggested method is accurate for this types of integral equations.

In example 3, three choices of M_1, M_2, N_1 and N_2 are compared. In case (I) we choose $M_1 = M_2 = 3$ and $N_1 = N_2 = 3$. In case (II) the values of N_1 and N_2 are increased by one without changing M_1 and M_2. Instead of case (I), the values of M_1 and M_2 are increased by one in case (III). The comparison of numerical results in cases (I) and (III) shows that increasing number of Chebyshev polynomials in each subregions doesn't lead to lower errors. It may happen due to errors obtained from computing the elements of matrix $K$and vector G in (21) by quadrature rules, or other computational errors. But, in case (II), increasing number of subregions obtained lower error than case (I). The fact that the dimensions of algebraic systems in case (II) and (III) are equal, illustrates the efficiency of method.

As shown in the examples, the powerful properties of Chebyshev polynomials, and simplicity of block-pulse functions, obtained more accurate solutions. Since the setting up of algebraic system (21) is simple, the suggested method can be used in practical cases easily. Furthermore, the current method can be run with increasing nubmer of parameters until the results settle down to an appropriate accuracy.

REFERENCES

1. Alipanah A, Esmaeili Sh. Numerical solution of the two-dimensional Fredholm integral equations using Gaussian radial basis function. J Comput Appl Math 2011;235:5342e7.
2. Atkinson KE. The numerical solutions of integral equations of the second kind. Cambridge: Cambridge University Press; 1997.
3. Babolian E, Maleknejad K, Roodaki M, Almasieh H. Two-dimensional triangular functions and its applications to nonlinear 2D Volterra-Fredholm integral equations. Comput Math Appl 2010;60:1711e22.
4. Chan R, Ng MK. Conjugate gradient methods for Toeplitz systems. SIAM Rev 1996;38:427e82.
5. Chihara TS. An introduction to orthogonal polynomials. New York: Gordon and Breach, Science Publisher Inc.; 1978.
6. Delves LM, Mohammed JL. Computational methods for integral equations. Cambridge: Cambridge University Press; 1985.
7. Guoqiang H, Jiong W. Extrapolation of Nystr€m solution for two dimensional nonlinear Fredholm integral equations. J Comput Appl Math 2001;134:259e68.
8. Ganti PR. Piecewise constant orthogonal functions and their application to systems and control. Springer-Verlag; 1983.
9. Hadizadeh M, Asgari M. An effective numerical approximation for the linear class of mixed integral equations. Appl Math Comput 2005;167:1090e100.

10. Hanson R, Phillips J. Numerical solution of two-dimensional integral equations using linear elements. SIAM J Numer Anal 1978;15:113e21.
11. Kadem A, Baleanu D. Two-dimensional transport equation as Fredholm integral equation. Commun Nonlinear Sci Numer Simulat 2012;17:530e5.
12. Kress R. Linear integral equations. New York: Springer-Verlag; 1989.
13. Liang F, Lin FR. A fast numerical solution method for two dimensional Fredholm integral equations of the second kind based on piecewise polynomial interpolation. Appl Math Comput 2010;216:3073e88.
14. Maleknejad K, Mahdiani K. Solving nonlinear mixed Volterra-Fredholm integral equations with two dimensional block- pulse functions using direct method. Commun Nonlinear Sci Numer Simulat 2011;16:3512e9.
15. Maleknejad K, Sohrabi S, Baranji B. Application of 2D-BPFs to nonlinear integral equations. Commun Nonlinear Sci Numer Simulat 2010;15:527e35.
16. Maleknejad K, JafariBehbahani Z. Applications of two-dimensional triangular functions for solving nonlinear class of mixed Volterra-Fredholm integral equations. Math Comput Model 2012;55:1833e44.
17. Rajan D, Chaudhuri S. Simultaneous estimation of super-resolved scene and depthmap from low resolution defocused observations. IEEE Trans Pattern Anal Mach Intell 2003;25:1102e17.
18. Tavassoli Kajani M, Hadi Vencheh A. Solving second kind integral equations with hybrid Chebyshev and Block-Pulse functions. Appl Math Comput 2005;163:71e7.
19. Volakis JL, Sertel K. Integral equation methods for electromagnetic. Scitech Publishing; 2012.

20. Xie WJ, Lin FR. A fast numerical solution method for two dimensional Fredholm integral equations of the second kind. Appl Numer Math 2009;59:1709e19.

CHAPTER 4

Calculus of variations with fractional derivatives and fractional integrals

Ricardo Almeida , Delfim F.M. Torres

Department of Mathematics, University of Aveiro, 3810-193 Aveiro, Portugal

ABSTRACT

We prove the Euler–Lagrange fractional equations and the sufficient optimality conditions for problems of the calculus of variations with functionals containing both fractional derivatives and fractional integrals in the sense of Riemann–Liouville.

KEYWORDS

Euler–Lagrange equation; Riemann–Liouville fractional derivative; Riemann–Liouville fractional integral

1. INTRODUCTION

In recent years numerous works have been dedicated to the fractional calculus of variations. Most of them deal with Riemann–Liouville fractional derivatives (see [1], [2],[3], [4] and [5] and the references therein), a few with Caputo or Riesz derivatives [6], [7],[8] and [9]. Depending on the type of the functional being considered, different fractional Euler–Lagrange type equations are obtained. We also mention [10], where a fractional Euler–Lagrange equation is obtained corresponding to a prescribed fractional space. Here we propose a new kind of functional with a Lagrangian containing not only a Riemann–Liouville fractional derivative (RLFD) but also a Riemann–Liouville fractional integral (RLFI). We prove the necessary conditions of Euler–Lagrange type for the fundamental fractional problem of the calculus of variations and for the fractional isoperimetric problem. Sufficient optimality conditions are also obtained under appropriate convexity assumptions.

2. FRACTIONAL CALCULUS

Fractional calculus is an interdisciplinary area, with many applications in several fields, such as engineering [11], [12] and [13], chemistry [14] and [15], electrical and electromechanical systems [16] and [17], viscoplasticity [18], physics [19], etc. [20].

In this section we review the necessary definitions and facts from fractional calculus. For more on the subject we refer the reader to [21], [22], [23] and [24].

Let f be a function defined on the interval $[a,b]$. Let α be a positive real and $n:=[\alpha]+1$.

DEFINITION 2.1. The left RLFI is defined by

$$ {}_aI_x^{\alpha} f(x) = \frac{1}{\Gamma(\alpha)} \int_a^x (x-t)^{\alpha-1} f(t)\mathrm{d}t, $$

and the right RLFI by

$$ {}_xI_b^{\alpha} f(x) = \frac{1}{\Gamma(\alpha)} \int_x^b (t-x)^{\alpha-1} f(t)\mathrm{d}t. $$

The left RLFD is defined by

$$ {}_aD_x^{\alpha} f(x) = \frac{d^n}{\mathrm{d}x^n} {}_aI_x^{n-\alpha} f(x) = \frac{1}{\Gamma(n-\alpha)} \frac{d^n}{\mathrm{d}x^n} \int_a^x (x-t)^{n-\alpha-1} f(t)\mathrm{d}t, $$

while the right RLFD is given by

$$ {}_xD_b^{\alpha} f(x) = (-1)^n \frac{d^n}{\mathrm{d}x^n} {}_xI_b^{n-\alpha} f(x) = \frac{(-1)^n}{\Gamma(n-\alpha)} \frac{d^n}{\mathrm{d}x^n} \int_x^b (t-x)^{n-\alpha-1} f(t)\mathrm{d}t. $$

The operators of Definition 2.1 are obviously linear. We now present the rules of fractional integration by parts for RLFI and RLFD. Let $p\geq 1, q\geq 1$, and $1/p+1/q\leq 1+\alpha$. If $g\in L_p(a,b)$ and $f\in L_q(a,b)$, then

$$ \int_a^b g(x)\, {}_aI_x^{\alpha} f(x)\mathrm{d}x = \int_a^b f(x)\, {}_xI_b^{\alpha} g(x)\mathrm{d}x; $$

if f, g, and the fractional derivatives ${}_aD_x^\alpha g$ and ${}_xD_b^\alpha f$ are continuous on $[a,b]$, then

$$\int_a^b g(x)\,{}_aD_x^\alpha f(x)\mathrm{d}x = \int_a^b f(x)\,{}_xD_b^\alpha g(x)\mathrm{d}x, \quad 0 < \alpha < 1.$$

REMARK 1. The left RLFD of f is infinite at $x=a$ if $f(a)\neq 0$ (cf. [25]). Similarly, the right RLFD is infinite if $f(b)\neq 0$. Thus, assuming that f possesses continuous left and right RLFD on $[a,b]$, then $f(a)=f(b)=0$ must be satisfied.

3. THE EULER–LAGRANGE EQUATION

Let us consider the following problem:

$$\mathcal{J}(y) = \int_a^b L(x, {}_aI_x^{1-\alpha}y(x), {}_aD_x^\beta y(x))\mathrm{d}x \longrightarrow \min. \tag{1}$$

We assume that $L(\cdot,\cdot,\cdot)\in C^1([a,b]\times R^2;R)$, $x\to\partial_2 L(x, {}_aI_x^{1-\alpha}y(x), {}_aD_x^\beta y(x))$ has continuous right RLFI of order $1-\alpha$ and $x\to\partial_3 L(x, {}_aI_x^{1-\alpha}y(x), {}_aD_x^\beta y(x))$ has continuous right RLFD of order β, where α and β are real numbers in the interval $(0,1)$.

REMARK 2. We are assuming that the admissible functions y are such that ${}_aI_x^{1-\alpha}y(x)$ and ${}_aD_x^\beta y(x)$ exist on the closed interval $[a,b]$. We also note that as α and β goes to 1 our fractional functional J tends to the classical functional $\int_a^b L(x, y(x), y'(x))\mathrm{d}x$ of the calculus of variations.

REMARK 3. We consider functionals J containing the left RLFI and the left RLFD only. This comprise the important cases in applications. The results of the paper are easily generalized for functionals containing also the right RLFI and/or right RLFD.

THEOREM 3.1 THE FRACTIONAL EULER–LAGRANGE EQUATION.

Let $y(\cdot)$ *be a local minimizer of problem* (1)*. Then,* $y(\cdot)$ *satisfies the fractional Euler–Lagrange equation*

$$ {}_xI_b^{1-\alpha}\,\partial_2 L(x, {}_aI_x^{1-\alpha}y(x), {}_aD_x^{\beta}y(x)) + {}_xD_b^{\beta}\,\partial_3 L(x, {}_aI_x^{1-\alpha}y(x), {}_aD_x^{\beta}y(x)) = 0 \quad \textit{for all } x \in [a,b]. \tag{2}$$

REMARK 4. Condition (2) is only necessary for an extremum. The question of sufficient conditions for an extremum is considered in Section 6.

PROOF. Since y is an extremizer of J, by a well known result of the calculus of variations the first variation of $J(\cdot)$ is zero at y, i.e.,

$$0 = \delta\mathcal{J}(\eta, y) = \int_a^b ({}_aI_x^{1-\alpha}\eta\,\partial_2 L + {}_aD_x^{\beta}\eta\,\partial_3 L)\mathrm{d}x. \tag{3}$$

Integrating by parts,

$$\int_a^b {}_aI_x^{1-\alpha}\eta\,\partial_2 L\mathrm{d}x = \int_a^b \eta\,{}_xI_b^{1-\alpha}\partial_2 L\mathrm{d}x \tag{4}$$

and

$$\int_a^b {}_aD_x^\beta \eta\, \partial_3 L \,\mathrm{d}x = \int_a^b \eta\, {}_xD_b^\beta \partial_3 L \,\mathrm{d}x. \tag{5}$$

Substituting (4) and (5) into Eq. (3), we find that $\int_a^b ({}_xI_b^{1-\alpha} \partial_2 L + {}_xD_b^\beta \partial_3 L)\eta \mathrm{d}x = 0$ for each η. Since η is an arbitrary function, by the fundamental lemma of the calculus of variations we deduce that ${}_xI_b^{1-\alpha} \partial_2 L + {}_xD_b^\beta \partial_3 L = 0$.

REMARK 5. As α and β goes to 1, the fractional Euler–Lagrange equations (2) becomes the classical Euler–Lagrange equation $\partial_2 L - d/\mathrm{d}x\, \partial_3 L = 0$.

A curve that is a solution of the fractional differential equation (2) will be called an*extremal* of J. Extremals play also an important role in the solution of the fractional isoperimetric problem (see Section 5). We note that Eq. (2) contains right RLFI and right RLFD, which are not present in the formulation of problem (1).

4. SOME GENERALIZATIONS

We now give some generalizations of Theorem 3.1.

4.1. Extension to variational problems of non-commensurate order

We now consider problems of the calculus of variations with Riemann–Liouville derivatives and integrals of non-commensurate order, i.e., we consider functionals containing RLFI and RLFD of different fractional orders. Let

$$\mathcal{J}(y) = \int_a^b L(x, {}_aI_x^{1-\alpha_1} y(x), \ldots, {}_aI_x^{1-\alpha_n} y(x), {}_aD_x^{\beta_1} y(x), \ldots, {}_aD_x^{\beta_m} y(x))\mathrm{d}x, \tag{6}$$

where n and m are two positive integers and $\alpha_i, \beta_j \in (0,1)$, $i=1,\ldots,n$ and $j=1,\ldots,m$. Following the proof of Theorem 3.1, we deduce the following result.

THEOREM 4.1. *If* $y(\cdot)$ *is a local minimizer of* (6), *then* $y(\cdot)$ *satisfies the Euler–Lagrange equation*

$$\sum_{i=1}^{n} {}_xI_b^{1-\alpha_i} \partial_{i+1} L + \sum_{j=1}^{m} {}_xD_b^{\beta_j} \partial_{j+n+1} L = 0 \quad \textit{for all } x \in [a, b].$$

4.2. Extension to several dependent variables

We now study the case of multiple unknown functions $y_1, \ldots, y_n$.

THEOREM 4.2. *Let* J *be the functional given by the expression*

$$\mathcal{J}(y_1, \ldots, y_n) = \int_a^b L(x, {}_aI_x^{1-\alpha} y_1(x), \ldots, {}_aI_x^{1-\alpha} y_n(x), {}_aD_x^{\beta} y_1(x), \ldots, {}_aD_x^{\beta} y_n(x))\mathrm{d}x.$$

If $y_1(\cdot), \ldots, y_n(\cdot)$ *is a local minimizer of* J, *then it satisfies for all* $x \in [a,b]$ *the following system of* n *fractional differential equations:*

$${}_xI_b^{1-\alpha} \partial_{k+1} L + {}_xD_b^{\beta} \partial_{n+k+1} L = 0, \quad k = 1, \ldots, n.$$

PROOF.

Denote by y and η the vectors $(y_1,\ldots,y_n)$ and $(\eta_1,\ldots,\eta_n)$, respectively. For a parameter ϵ, we consider a new function

$$J(\epsilon)=J(y+\epsilon\eta). \tag{7}$$

Since $y_1(\cdot),\ldots,y_n(\cdot)$ is an extremizer of J, $J'(0)=0$. Differentiating Eq. (7) with respect to ϵ, at $\epsilon=0$, we obtain

$$\int_a^b \left[{}_aI_x^{1-\alpha}\eta_1\,\partial_2 L+\cdots+{}_aI_x^{1-\alpha}\eta_n\,\partial_{n+1}L+{}_aD_x^{\beta}\eta_1\,\partial_{n+2}L+\cdots+{}_aD_x^{\beta}\eta_n\,\partial_{2n+1}L\right]\mathrm{d}x=0.$$

Integrating by parts leads to

$$\int_a^b \left[{}_xI_b^{1-\alpha}\partial_2 L+{}_xD_b^{\beta}\partial_{n+2}L\right]\eta_1+\cdots+\left[{}_xI_b^{1-\alpha}\partial_{n+1}L+{}_xD_b^{\beta}\partial_{2n+1}L\right]\eta_n\,\mathrm{d}x=0.$$

Consider a variation $\eta=(\eta_1,0,\ldots,0)$, η_1 arbitrary; then by the fundamental lemma of the calculus of variations we obtain ${}_xI_b^{1-\alpha}\partial_2 L+{}_xD_b^{\beta}\partial_{n+2}L=0$. Selecting appropriate variations η, one deduce the remaining formulas.

5. THE FRACTIONAL ISOPERIMETRIC PROBLEM

We consider now the problem of minimizing the functional J given by (1) subject to an integral constraint $\mathcal{I}(y)=\int_a^b g(x,{}_aI_x^{1-\alpha}y(x),{}_aD_x^{\beta}y(x))\mathrm{d}x=l$, where l is a prescribed value. This problem was solved in [8] for functionals containing Caputo fractional derivatives and RLFI. Using similar techniques as the ones discussed in [8], one proves the following:

THEOREM 5.1. *Consider the problem of minimizing the functional* J *as in* (1)*on the set of functions* y *satisfying condition* I(y)=l. *Let* y *be a local minimum for the problem. Then, there exist two constants* λ_0 *and* λ, *not both zero, such that* y *satisfies the Euler–Lagrange equation* ${}_xI_b^{1-\alpha}\partial_2 K + {}_xD_b^{\beta}\partial_3 K = 0$ *for all* x∈[a,b], *where* $K=\lambda_0 L+\lambda g$.

REMARK 6. If y is not an extremal for I, then one can choose $\lambda_0=1$ in Theorem 5.1: there exists a constant λ such that y satisfies ${}_xI_b^{1-\alpha}\partial_2 F + {}_xD_b^{\beta}\partial_3 F = 0$ for all x∈[a,b], where F=L+λg.

6. SUFFICIENT CONDITIONS

In this section we prove the sufficient conditions that ensure the existence of minimums. Similarly to what happens in the classical calculus of variations, some conditions of convexity are in order.

DEFINITION 6.1. Given a function L, we say that $L(\underline{x}, u, v)$ is convex in $S \subseteq R^3$ if $\partial_2 L$ and $\partial_3 L$ exist and are continuous and verify the following condition:

$$L(x,u+u_1,v+v_1)-L(x,u,v) \geq \partial_2 L(x,u,v)u_1+\partial_3 L(x,u,v)v_1$$

for all $(x,u,v),(x,u+u_1,v+v_1) \in S$.

Similarly, we define convexity for $L(\underline{x}, \underline{u}, v)$.

THEOREM 6.2. *Let* $L(\underline{x}, u, v)$ *be a convex function in* $[a,b]\times R^2$ *and let* y_0 *be a curve satisfying the fractional Euler–Lagrange equation* (2). *Then,* y_0 *minimizes* (1).

PROOF. The following holds:

$$\begin{aligned}\mathcal{J}(y_0+\eta)-\mathcal{J}(y_0) &= \int_a^b \left[L(x, {}_aI_x^{1-\alpha} y_0(x) + {}_aI_x^{1-\alpha}\eta(x), {}_aD_x^{\beta} y_0(x) + {}_aD_x^{\beta}\eta(x)) - L(x, {}_aI_x^{1-\alpha} y_0(x), {}_aD_x^{\beta} y_0(x))\right] dx \\ &\geq \int_a^b \left[\partial_2 L(x, {}_aI_x^{1-\alpha} y_0(x), {}_aD_x^{\beta} y_0(x))\, {}_aI_x^{1-\alpha}\eta + \partial_3 L(x, {}_aI_x^{1-\alpha} y_0(x), {}_aD_x^{\beta} y_0(x))\, {}_aD_x^{\beta}\eta\right] dx \\ &= \int_a^b \left[{}_xI_b^{1-\alpha}\partial_2 L + {}_xD_b^{\beta}\partial_3 L\right]_{(x, {}_aI_x^{1-\alpha} y_0(x), {}_aD_x^{\beta} y_0(x))} \eta\, dx = 0.\end{aligned}$$

Thus, $J(y_0+\eta) \geq J(y_0)$.

We now present a sufficient condition for convex Lagrangians on the third variable only. First we recall the notion of exact field.

DEFINITION 6.3. Let $D \subseteq R^2$ and let $\Phi : D \to R$ be a function of class C^1. We say that Φ is an exact field for L covering D if there exists a function $S \in C^1(D, R)$ such that

$$\partial_1 S(x,y) = L(x,y,\Phi(x,y)) - \partial_3 L(x,y,\Phi(x,y))\Phi(x,y),$$

$$\partial_2 S(x,y) = \partial_3 L(x,y,\Phi(x,y)).$$

REMARK 7. This definition is motivated by the classical Euler–Lagrange equation. Indeed, every solution $y_0 \in C^2[a,b]$ of the differential equation $y' = \Phi(x, y(x))$ satisfies the (classical) Euler–Lagrange equation $\partial_2 L - \frac{d}{dx}\partial_3 L = 0$.

THEOREM 6.4. *Let* $L(x, u, v)$ *be a convex function in* $[a,b] \times R^2$, Φ *an exact field for* L *covering* $[a,b] \times R \subseteq D$, *and* y_0 *a solution of the fractional equation*

equation(8)

$${}_aD_x^{\alpha} y(x) = \Phi(x, {}_aI_x^{1-\alpha} y(x)). \qquad (8)$$

Then, y_0 *is a minimizer for* $\mathcal{J}(y) = \int_a^b L(x, {}_aI_x^{1-\alpha}y(x), {}_aD_x^{\alpha}y(x))dx$ *subject to the constraint*

$$\left\{y : [a,b] \to \mathbb{R} \mid {}_aI_a^{1-\alpha}y(a) = {}_aI_a^{1-\alpha}y_0(a), {}_aI_b^{1-\alpha}y(b) = {}_aI_b^{1-\alpha}y_0(b)\right\}. \quad (9)$$

PROOF. Let $E(x,y,z,w)=L(x,y,w)-L(x,y,z)-\partial_3L(x,y,z)(w-z)$. First observe that

$$\begin{aligned}\frac{d}{dx}S(x, {}_aI_x^{1-\alpha}y(x)) &= \partial_1 S(x, {}_aI_x^{1-\alpha}y(x)) + \partial_2 S(x, {}_aI_x^{1-\alpha}y(x))\frac{d}{dx}{}_aI_x^{1-\alpha}y(x)\\ &= \partial_1 S(x, {}_aI_x^{1-\alpha}y(x)) + \partial_2 S(x, {}_aI_x^{1-\alpha}y(x)){}_aD_x^{\alpha}y(x).\end{aligned}$$

Since $E \geq 0$, it follows that

$$\begin{aligned}\mathcal{J}(y) &= \int_a^b \Big[E(x, {}_aI_x^{1-\alpha}y, \Phi(x, {}_aI_x^{1-\alpha}y), {}_aD_x^{\alpha}y) + L(x, {}_aI_x^{1-\alpha}y, \Phi(x, {}_aI_x^{1-\alpha}y))\\ &\quad + \partial_3 L(x, {}_aI_x^{1-\alpha}y, \Phi(x, {}_aI_x^{1-\alpha}y))({}_aD_x^{\alpha}y - \Phi(x, {}_aI_x^{1-\alpha}y))\Big]dx\\ &\geq \int_a^b \Big[L(x, {}_aI_x^{1-\alpha}y, \Phi(x, {}_aI_x^{1-\alpha}y)) + \partial_3 L(x, {}_aI_x^{1-\alpha}y, \Phi(x, {}_aI_x^{1-\alpha}y))({}_aD_x^{\alpha}y - \Phi(x, {}_aI_x^{1-\alpha}y))\Big]dx\\ &= \int_a^b \Big[\partial_1 S(x, {}_aI_x^{1-\alpha}y) + \partial_2 S(x, {}_aI_x^{1-\alpha}y){}_aD_x^{\alpha}y\Big]dx\\ &= \int_a^b \frac{d}{dx}S(x, {}_aI_x^{1-\alpha}y)dx\\ &= S(b, {}_aI_b^{1-\alpha}y(b)) - S(a, {}_aI_a^{1-\alpha}y(a)).\end{aligned}$$

Because y_0 is a solution of (8), $E(x, {}_aI_x^{1-\alpha}y_0, \Phi(x, {}_aI_x^{1-\alpha}y_0), {}_aD_x^{\alpha}y_0) = 0$. With similar calculations as before, one has $\mathcal{J}(y_0) = S(b, {}_aI_b^{1-\alpha}y_0(b)) - S(a, {}_aI_a^{1-\alpha}y_0(a))$. We just proved that $J(y_0) \leq J(y)$ when subject to the constraint (9).

7. CONCLUSIONS

In this note we consider a new class of fractional functionals of the calculus of variations that depend not only on fractional derivatives but also on fractional integrals. We exhibit necessary and sufficient conditions of optimality for the fundamental problem of the calculus of variations and for problems subject to integral constrains (isoperimetric problems). As future work it would be interesting to address the question of existence of solutions, and to study direct methods to minimize the proposed type of functionals.

ACKNOWLEDGMENTS

The authors are grateful to the support of the *Control Theory Group* from the *Centre for Research on Optimization and Control* given by the *Portuguese Foundation for Science and Technology* (FCT), cofinanced through the *European Community Fund*FEDER/POCI 2010.

REFERENCES

1. T.M. Atanacković, S. Konjik, S. Pilipović, Variational problems with fractional derivatives: Euler Lagrange equations, J. Phys. A: Math. Theor. 41 (9) (2008) 095201.
2. D. Baleanu, S.I. Muslih, E.M. Rabei, On fractional Euler–Lagrange and Hamilton equations and the fractional generalization of total time derivative, Nonlinear Dynam. 53 (1–2) (2008) 67–74.
3. R.A. El-Nabulsi, D.F.M. Torres, Necessary optimality conditions for fractional action-like integrals of variational calculus with Riemann–Liouville derivatives of order (α, β), Math. Methods Appl. Sci. 30 (15) (2007) 1931–1939.
4. G.S.F. Frederico, D.F.M. Torres, A formulation of Noether's theorem for fractional problems of the calculus of variations, J. Math. Anal. Appl. 334 (2)(2007) 834–846.

5. G.S.F. Frederico, D.F.M. Torres, Fractional conservation laws in optimal control theory, Nonlinear Dynam. 53 (3) (2008) 215–222.
6. O.P. Agrawal, Fractional variational calculus and the transversality conditions, J. Phys. A: Math. Gen. 39 (2006) 10375–10384.
7. O.P. Agrawal, Fractional variational calculus in terms of Riesz fractional derivatives, J. Phys. A 40 (24) (2007) 6287–6303.
8. R. Almeida, D.F.M. Torres, Necessary and sufficient conditions for the fractional calculus of variations with Caputo derivatives (submitted for publication).
9. D. Baleanu, Om.P. Agrawal, Fractional Hamilton formalism within Caputo's derivative, Czechoslovak J. Phys. 56 (10–11) (2006) 1087–1092.
10. S.I. Muslih, D. Baleanu, Fractional Euler–Lagrange equations of motion in fractional space, J. Vib. Control 13 (9–10) (2007) 1209–1216.
11. N. Ferreira, F. Duarte, M. Lima, M. Marcos, J.T. Machado, Application of fractional calculus in the dynamical analysis and control of mechanical manipulators, Fract. Calc. Appl. Anal. 11 (1) (2008) 91–113.
12. V.V. Kulish, J.L. Lage, Application of fractional calculus to fluid mechanics, J. Fluids Eng. 124 (3) (2002) 803–806.
13. R. Magin, Fractional calculus in Bioengineering. Part 1–3, Critical Reviews in Bioengineering 32 (2004).
14. R. Hilfer, Fractional diffusion based on Riemann–Liouville fractional derivatives, J. Phys. Chem. B 104 (16) (2000) 3914–3917.
15. F. Metzler, W. Schick, H.G. Kilian, T.F. Nonnenmacher, Relaxation in filled polymers: A fractional calculus approach, J. Chem. Phys. 103 (1995) 7180–7186.
16. L. Debnath, Recent applications of fractional calculus to science and engineering, Int. J. Math. Math. Sci. 54 (2003) 3413–3442.

17. A. Oustaloup, V. Pommier, P. Lanusse, Design of a fractional control using performance contours. Application to an electromechanical system, Fract.Calc. Appl. Anal. 6 (1) (2003) 1–24.
18. K. Diethelm, A.D. Freed, On the solution of nonlinear fractional order differential equations used in the modeling of viscoplasticity, in: F. Keil,
19. W. Mackens, H. Voss, J. Werther (Eds.), Scientific Computing in Chemical Engineering II-Computational Fluid Dynamics, Reaction Engineering and Molecular Properties, Springer-Verlag, Heidelberg, 1999, pp. 217–224.
20. R. Hilfer, Applications of Fractional Calculus in Physics, World Scientific, Singapore, 2000.
21. J.A. Tenreiro Machado, R.S. Barbosa, Introduction to the special issue on fractional differentiation and its applications, J. Vib. Control 9–10 (2008) 1253. [21] A.A. Kilbas, H.M. Srivastava, J.J. Trujillo, Theory and Applications of Fractional Differential Equations, Elsevier, Amsterdam, 2006.
22. K.B. Oldham, J. Spanier, The Fractional Calculus, Academic Press [A subsidiary of Harcourt Brace Jovanovich, Publishers], New York, 1974.
23. I. Podlubny, Fractional Differential Equations, Academic Press, Inc, San Diego, CA, 1999.
24. S.G. Samko, A.A. Kilbas, O.I. Marichev, Fractional Integrals and Derivatives, Gordon and Breach Science Publishers, Yverdon, 1993.
25. B. Ross, S.G. Samko, E.R. Love, Functions that have no first order derivative might have fractional derivatives of all orders less than one, Real Annal.Exchange 20 (1) (1994/95) 140–157.

CHAPTER 5

Numerical solution of two-dimensional fuzzy Fredholm integral equations of the second kind using triangular functions

Farshid Mirzaee[*], **Mohammad Komak Yari, Elham Hadadiyan**

Faculty of Mathematical Sciences and Statistics, Malayer University, P. O. Box 65719-95863, Malayer, Iran

ABSTRACT

The main purpose of this paper is to approximate the solution of linear two-dimensional fuzzy Fredholm integral equations of the second kind (2D-FFIE-2). We use fuzzy two-dimensional triangular functions (2D-TFs) to reduce the 2D–FFIE–2 to a system of linear Fredholm integral equations of the second kind with three variables in crisp. More over, we prove the convergence of the method. Finally we illustrate this method with some numerical examples to demonstrate the validity and applicability of the technique.

KEYWORDS

Fuzzy number; Two-dimensional fuzzy Fredholm integral equations; Two-dimensional orthogonal triangular functions

1. INTRODUCTION

Fuzzy integral equations are important in studying and solving a large proportion of the problems in many topics in applied mathematics, in particular in relation to physics, geographic, medical, biology. Usually in many applications some of the parameters in our problems are represented by fuzzy number rather than crisp, and hence it is important to develop mathematical models and numerical procedures that would appropriately treat general fuzzy integral equations and solve them.

Recently, Mirzaee and Bimesl (2014) adapted the matrix method for the Fredholm integral equations. The study of fuzzy integral equations (FIEs) which attracted growing interest for some time, begins with the investigations of Kaleva (1987) and Seikkala (1987) for the fuzzy Volterra integral equation that is equivalent to the initial value problem for first order fuzzy differential equations. These studies continued by Wang, 1984, Nanda, 1989, Ralescu and Adams, 1980, Bede and Gal, 2005 and Goetschel and Voxman, 1986 and others. In Wu (2000) investigated the fuzzy Riemann integral and its numerical integration. Molabahrami et al. (2011) have used the parametric form of a fuzzy number and they have converted a linear fuzzy Fredholm integral equation to two linear systems of integral equations of the second kind in the crisp case.

Recently, some numerical methods have been investigated to solve linear fuzzy Fredholm integral equations of the second kind in two-dimensional space. For example,Rivaz and Yousefi (2012) and Ezzati and Ziari (2013) used homotopy perturbation method and fuzzy Bivariate Bernestein polynomials method for solving 2D-FFIE-2, respectively. Deb et al. (2006) introduced a new set of orthogonal functions, a numerical scheme based on such functions was applied for solving variational problem and integral equation by Babolian et al., 2007, Babolian et al., 2009 and Babolian et al., 2010.

In this paper, we apply the triangular functions for approximate the solutions of the linear two-dimensional Fredholm fuzzy integral equations of the second kind for the first time.

This paper is organized as follows. In Section 2, we present some definitions and properties of one and two-dimensional triangular functions which will be used later. In Section 3, we give an overview of elementary concepts of the fuzzy calculus. two-dimensional fuzzy Fredholm integral equation is described in Section 4. In Section 5, we apply 2D-TFs for solving linear two-dimensional fuzzy Fredholm integral equation. Section 6 is concerned with discussing the convergency of the proposed method then this method is implemented for solving two illustrative examples in Section 7 and finally, conclusion is drawn in Section 8.

2. PRELIMINARIES

2.1. A review of one-dimensional triangular functions

DEFINITION 2.1. Two m-sets of triangular functions (TFs) are defined over the interval $[0,T]$ as:

$$T1_i(t) = \begin{cases} 1 - \dfrac{t - ih}{h} & ih \leq t \leq (i+1)h, \\ 0 & otherwise, \end{cases} \quad (1)$$

$$T2_i(t) = \begin{cases} \dfrac{t - ih}{h} & ih \leq t \leq (i+1)h, \\ 0 & otherwise, \end{cases} \quad (2)$$

where $i = 0,1,\ldots,m - 1$, $h = T/m$, with a positive integer value for m. We have

$$\int_0^1 T1_i(t)T1_j(t)dt = \int_0^1 T2_i(t)T2_j(t)dt = \begin{cases} \dfrac{h}{3} & i = j, \\ 0 & i \neq j, \end{cases} \quad (3)$$

and

$$\int_0^1 T1_i(t)T2_j(t)dt = \int_0^1 T2_i(t)T1_j(t)dt = \begin{cases} \dfrac{h}{6} & i = j, \\ 0 & i \neq j. \end{cases} \quad (4)$$

Also, consider $T1i$ as the i th left-handed triangular function and $T2i$ as the i th right-handed triangular function. In this paper, it is assumed that $T = 1$.

Consider the first m terms of the left-handed triangular functions and the first m terms of the right-handed triangular functions and write them concisely as m-vectors:

$$T1(t) = \left[T1_0(t), T1_1(t), \ldots, T1_{m-1}(t)\right]^T, \tag{5}$$

$$T2(t) = \left[T2_0(t), T2_1(t), \ldots, T2_{m-1}(t)\right]^T, \tag{6}$$

where $T1(t)$ and $T2(t)$ are called left-handed triangular functions ($LHTF$) vector and right-handed triangular functions ($RHTF$) vector, respectively. We have:

$$\int_0^1 T1(t)T1^T(t)dt = \int_0^1 T2(t)T2^T(t)dt = \frac{h}{3}I, \tag{7}$$

$$\int_0^1 T1(t)T2^T(t)dt = \int_0^1 T2(t)T1^T(t)dt = \frac{h}{6}I, \tag{8}$$

which I is an $m \times m$ identity matrix. We denote the 1D-TF vector $T(t)$ as follows

$$T(t) = \begin{bmatrix} T1(t) \\ T2(t) \end{bmatrix}. \tag{9}$$

The expansion of the function $f(t)$ over [0,1] with respect to 1D-TFs may be written as

$$f(t) \approx \sum_{i=0}^{m-1} c_i T1_i(t) + \sum_{i=0}^{m-1} d_i T2_i(t) = C^T \cdot T1(t) + D^T \cdot T2(t) = \begin{bmatrix} C \\ D \end{bmatrix}^T \cdot \begin{bmatrix} T1(t) \\ T2(t) \end{bmatrix} = F^T \cdot T(t),$$

where Ci and Di are samples of f, for example $Ci = f(ih)$ and $Di = f((i + 1)h)$ for $i = 0,1,\ldots,m - 1$, so there is no need for integration. The vector F is called the 1D-TF coefficient vector.

2.2. Two-dimensional triangular functions and their properties

An $(m_1 \times m_2)$-set of the region $(\Omega = [0,1] \times [0,1])$ is defined by

$$T_{i,j}^{1,1}(s,t) = \begin{cases} (1 - \frac{s - ih_1}{h_1})(1 - \frac{t - jh_2}{h_2}) & ih_1 \leq s \leq (i+1)h_1, \\ & jh_2 \leq t \leq (j+1)h_2, \\ 0 & otherwise, \end{cases} \tag{10}$$

$$T_{i,j}^{1,2}(s,t) = \begin{cases} (1 - \frac{s - ih_1}{h_1})(\frac{t - jh_2}{h_2}) & ih_1 \leq s \leq (i+1)h_1, \\ & jh_2 \leq t \leq (j+1)h_2, \\ 0 & otherwise, \end{cases} \tag{11}$$

$$T_{i,j}^{2,1}(s,t) = \begin{cases} (\frac{s - ih_1}{h_1})(1 - \frac{t - jh_2}{h_2}) & ih_1 \leq s \leq (i+1)h_1, \\ & jh_2 \leq t \leq (j+1)h_2, \\ 0 & otherwise, \end{cases} \tag{12}$$

$$T_{i,j}^{2,2}(s,t) = \begin{cases} (\frac{s - ih_1}{h_1})(\frac{t - jh_2}{h_2}) & ih_1 \leq s \leq (i+1)h_1, \\ & jh_2 \leq t \leq (j+1)h_2, \\ 0 & otherwise, \end{cases} \tag{13}$$

where $i = 0,1,\ldots,m_1 - 1$, $j = 0,1,\ldots,m_2 - 1$
and $h_1 = 1/m_1$, $h_2 = 1/m_2$. m_1 and m_2 are arbitrary positive integers. It is clear that

$$\begin{aligned} T_{i,j}^{1,1}(s,t) &= T1_i(s).T1_j(t), \\ T_{i,j}^{1,2}(s,t) &= T1_i(s).T2_j(t), \\ T_{i,j}^{2,1}(s,t) &= T2_i(s).T1_j(t), \\ T_{i,j}^{2,2}(s,t) &= T2_i(s).T2_j(t). \end{aligned} \tag{14}$$

Furthermore,

$$T_{i,j}^{1,1}(s,t) + T_{i,j}^{1,2}(s,t) + T_{i,j}^{2,1}(s,t) + T_{i,j}^{2,2}(s,t) = \Phi_{i,j}(s,t),$$

where $\Phi i,j(s,t)$ is the $\{im_2 + j + 1\}$ th block-pulse function defined on $ih_1 \leq s \leq (i + 1)h_1$ and $jh_2 \leq t \leq (j + 1)h_2$ as

$$\Phi_{i,j}(s,t) = \begin{cases} 1 & \begin{array}{l} ih_1 \leq s \leq (i+1)h_1, \\ jh_2 \leq t \leq (j+1)h_2, \end{array} \\ 0 & \quad otherwise. \end{cases}$$

From Eq.s (3), (4) and (14) we have:

$$\int_0^1 \int_0^1 T_{i_1,j_1}^{p_1,q_1}(s,t) \cdot T_{i_2,j_2}^{p_2,q_2}(s,t) ds dt = \Delta_{p_1,p_2} \delta_{i_1,i_2} \cdot \Delta_{q_1,q_2} \delta_{j_1,j_2},$$

where δ denotes the Kronecker delta function and

$$\Delta_{\alpha,\beta} = \begin{cases} \frac{h}{3} & \alpha = \beta \in \{1,2\}, \\ \frac{h}{6} & \quad \alpha \neq \beta. \end{cases}$$

On the other hand, if

$$T11(s,t) = \left[T_{0,0}^{1,1}(s,t), \ldots, T_{0,m_2-1}^{1,1}(s,t), T_{1,0}^{1,1}(s,t), \ldots, T_{m_1-1,m_2-1}^{1,1}(s,t)\right]^T,$$

$$T12(s,t) = \left[T_{0,0}^{1,2}(s,t), \ldots, T_{0,m_2-1}^{1,2}(s,t), T_{1,0}^{1,2}(s,t), \ldots, T_{m_1-1,m_2-1}^{1,2}(s,t)\right]^T,$$

$$T21(s,t) = \left[T_{0,0}^{2,1}(s,t), \ldots, T_{0,m_2-1}^{2,1}(s,t), T_{1,0}^{2,1}(s,t), \ldots, T_{m_1-1,m_2-1}^{2,1}(s,t)\right]^T,$$

$$T22(s,t) = \left[T_{0,0}^{2,2}(s,t), \ldots, T_{0,m_2-1}^{2,2}(s,t), T_{1,0}^{2,2}(s,t), \ldots, T_{m_1-1,m_2-1}^{2,2}(s,t)\right]^T,$$

then $T(s,t)$, the $2D$-TF vector, can be defined as

$$T(s,t) = \begin{bmatrix} T11(s,t) \\ T12(s,t) \\ T21(s,t) \\ T22(s,t) \end{bmatrix}_{4m_1m_2 \times 1} \quad (15)$$

We have:

$$\int_0^1\int_0^1 T11^T(s,t)T11(s,t)dsdt = \frac{h_1}{3}I_{m_1\times m_1}\otimes\frac{h_2}{3}I_{m_2\times m_2},$$

$$\int_0^1\int_0^1 T11^T(s,t)T12(s,t)dsdt = \frac{h_1}{3}I_{m_1\times m_1}\otimes\frac{h_2}{6}I_{m_2\times m_2},$$

$$\int_0^1\int_0^1 T11^T(s,t)T21(s,t)dsdt = \frac{h_1}{6}I_{m_1\times m_1}\otimes\frac{h_2}{3}I_{m_2\times m_2},$$

$$\int_0^1\int_0^1 T11^T(s,t)T22(s,t)dsdt = \frac{h_1}{6}I_{m_1\times m_1}\otimes\frac{h_2}{6}I_{m_2\times m_2},$$

where ⊗ denotes the Kronecker product defined for two arbitrary matrices *P* and *Q* as

$P\otimes Q = p_{i,j}Q.$

The same equations are implied for *T*12(*s*,*t*), *T*21(*s*,*t*) and *T*22(*s*,*t*), by similar computations. Hence, we can carry out double integration of *T*(*s*,*t*):

equation(16)

$$\int_0^1\int_0^1 T^T(s,t)T(s,t)dtds = D, \tag{16}$$

which *D* is $(4m_1m_2 \times 4m_1m_2)$-matrix as follows:

$$D=\begin{bmatrix} \frac{h_1}{3}I_1\otimes\frac{h_2}{3}I_2 & \frac{h_1}{3}I_1\otimes\frac{h_2}{6}I_2 & \frac{h_1}{6}I_1\otimes\frac{h_2}{3}I_2 & \frac{h_1}{6}I_1\otimes\frac{h_2}{6}I_2 \\ \frac{h_1}{3}I_1\otimes\frac{h_2}{6}I_2 & \frac{h_1}{3}I_1\otimes\frac{h_2}{3}I_2 & \frac{h_1}{6}I_1\otimes\frac{h_2}{6}I_2 & \frac{h_1}{6}I_1\otimes\frac{h_2}{3}I_2 \\ \frac{h_1}{6}I_1\otimes\frac{h_2}{3}I_2 & \frac{h_1}{6}I_1\otimes\frac{h_2}{6}I_2 & \frac{h_1}{3}I_1\otimes\frac{h_2}{3}I_2 & \frac{h_1}{3}I_1\otimes\frac{h_2}{6}I_2 \\ \frac{h_1}{6}I_1\otimes\frac{h_2}{6}I_2 & \frac{h_1}{6}I_1\otimes\frac{h_2}{3}I_2 & \frac{h_1}{3}I_1\otimes\frac{h_2}{6}I_2 & \frac{h_1}{3}I_1\otimes\frac{h_2}{3}I_2 \end{bmatrix}, \quad (17)$$

which $I_1 = I_{m1\times m1}$ and $I_2 = I_{m2\times m2}$.

2.3. Function expansion with 2*D*-TFs

We can approximate the function $f(s,t)$ defined over Ω by 2D-TFs as follows

$$\begin{aligned} f(s,t) &\approx \sum_{i=0}^{m_1-1}\sum_{j=0}^{m_2-1} c_{i,j}T_{i,j}^{1,1}(s,t) + \sum_{i=0}^{m_1-1}\sum_{j=0}^{m_2-1} d_{i,j}T_{i,j}^{1,2}(s,t) \\ &\quad + \sum_{i=0}^{m_1-1}\sum_{j=0}^{m_2-1} e_{i,j}T_{i,j}^{2,1}(s,t) + \sum_{i=0}^{m_1-1}\sum_{j=0}^{m_2-1} l_{i,j}T_{i,j}^{2,2}(s,t) \\ &= C^T\cdot T11(s,t) + D^T\cdot T12(s,t) + E^T\cdot T21(s,t) \\ &\quad + L^T\cdot T22(s,t) = F^T\cdot T(s,t), \end{aligned}$$

where F is a $4m_1m_2$-vector given by

$$F = [C^T \quad D^T \quad E^T \quad L^T]^T,$$

and $T(s,t)$ is defined in Eq. (15). The 2*D*-TF coefficients in C, D, E and L can be computed by sampling the function $f(s,t)$ at grid points si and tj such that $si = ih_1$ and$tj = jh_2$, for various i and j. So we have

$$C_k = c_{i,j} = f(s_i, t_j),$$

$$D_k = d_{i,j} = f(s_i, t_{j+1}),$$

$$E_k = e_{i,j} = f(s_{i+1}, t_j),$$

$$L_k = l_{i,j} = f(s_{i+1}, t_{j+1}),$$

where $k = im_2 + j$ and $i = 0,1,\ldots,m_1 - 1, j = 0,1,\ldots,m_2 - 1$. The $4m_1m_2$-vector F is called the $2D$-TF coefficient vector.

Let $u(s,t,r)$ be a function of three variables on $(\Omega \times [0,1])$. It can be approximated with respect to $2D$-TFs and 1D-TFs as follows:

$$u(s,t,r) \approx T^{T}(s,t) \cdot U \cdot T(r), \tag{18}$$

where $T(s,t)$ and $T(r)$ are $2D$-TF vector and 1D-TF vector of dimension $4m_1m_2$ and $2m_3$, respectively and U is a $(4m_1m_2 \times 2m_3)$ $2D$-TF coefficient matrix. This matrix can be represented as

$$U = \begin{bmatrix} U11 & U12 \\ U21 & U22 \\ U31 & U32 \\ U41 & U42 \end{bmatrix},$$

where each block of U is an $(m_1m_2 \times m_3)$-matrix that can be computed by sampling the function $u(s,t,r)$ at grid points (si,tj,rk) such that

$$s_i = ih_1, \quad i = 0,1,\ldots,m_1 - 1, \quad h_1 = \frac{1}{m_1},$$

$$t_j = jh_2, \quad j = 0,1,\ldots,m_2 - 1, \quad h_2 = \frac{1}{m_2},$$

$$r_k = kh_3, \quad k = 0,1,\ldots,m_3 - 1, \quad h_3 = \frac{1}{m_3}.$$

Let $l = im_2 + j$, then

$$U11_{l,k} = u(s_i, t_j, r_k),$$
$$U12_{l,k} = u(s_i, t_j, r_{k+1}),$$
$$U21_{l,k} = u(s_i, t_{j+1}, r_k),$$
$$U22_{l,k} = u(s_i, t_{j+1}, r_{k+1}),$$

$$U31_{l,k} = u(s_{i+1}, t_j, r_k),$$
$$U32_{l,k} = u(s_{i+1}, t_j, r_{k+1}),$$
$$U41_{l,k} = u(s_{i+1}, t_{j+1}, r_k),$$
$$U42_{l,k} = u(s_{i+1}, t_{j+1}, r_{k+1}).$$

Let $k(s,t,x,y)$ be a function of four variables on $(\Omega \times \Omega)$. It can be approximated with respect to $2D$-TFs as follows:

$$k(s,t,x,y) \approx T^T(s,t) \cdot K \cdot T(x,y), \tag{19}$$

where $T(s,t)$ and $T(x,y)$ are $2D$-TF vectors of dimension $4m_1m_2$ and $4m_3m_4$, respectively and K is a $(4m_1m_2 \times 4m_3m_4)$ $2D$-TF coefficient matrix. This matrix can be represented as

$$K = \begin{bmatrix} K11 & K12 & K13 & K14 \\ K21 & K22 & K23 & K24 \\ K31 & K32 & K33 & K34 \\ K41 & K42 & K43 & K44 \end{bmatrix}, \tag{20}$$

where each block of K is an $(m_1m_2 \times m_3m_4)$-matrix that can be computed by sampling the function $k(s,t,x,y)$ at grid points $(s_{i1}, t_{j1}, x_{i2}, y_{j2})$ such that

$$s_{i_1} = i_1 h_1, \quad i_1 = 0, 1, \ldots, m_1 - 1, \quad h_1 = \frac{1}{m_1},$$

$$t_{j_1} = j_1 h_2, \quad j_1 = 0, 1, \ldots, m_2 - 1, \quad h_2 = \frac{1}{m_2},$$

$$x_{i_2} = i_2 h_3, \quad i_2 = 0, 1, \ldots, m_3 - 1, \quad h_3 = \frac{1}{m_3},$$

$$y_{j_2} = j_2 h_4, \quad j_2 = 0, 1, \ldots, m_4 - 1, \quad h_4 = \frac{1}{m_4}.$$

Let $p = i_1 m_2 + j_1$ and $q = i_2 m_4 + j_2$, then

$$K11_{p,q} = k(s_{i_1}, t_{j_1}, x_{i_2}, y_{j_2}),$$
$$K12_{p,q} = k(s_{i_1}, t_{j_1}, x_{i_2}, y_{j_2+1}),$$
$$K13_{p,q} = k(s_{i_1}, t_{j_1}, x_{i_2+1}, y_{j_2}),$$
$$K14_{p,q} = k(s_{i_1}, t_{j_1}, x_{i_2+1}, y_{j_2+1}),$$

$$K21_{p,q} = k(s_{i_1}, t_{j_1+1}, x_{i_2}, y_{j_2}),$$
$$K22_{p,q} = k(s_{i_1}, t_{j_1+1}, x_{i_2}, y_{j_2+1}),$$
$$K23_{p,q} = k(s_{i_1}, t_{j_1+1}, x_{i_2+1}, y_{j_2}),$$
$$K24_{p,q} = k(s_{i_1}, t_{j_1+1}, x_{i_2+1}, y_{j_2+1}),$$

$$K31_{p,q} = k(s_{i_1+1}, t_{j_1}, x_{i_2}, y_{j_2}),$$
$$K32_{p,q} = k(s_{i_1+1}, t_{j_1}, x_{i_2}, y_{j_2+1}),$$
$$K33_{p,q} = k(s_{i_1+1}, t_{j_1}, x_{i_2+1}, y_{j_2}),$$
$$K34_{p,q} = k(s_{i_1+1}, t_{j_1}, x_{i_2+1}, y_{j_2+1}),$$

$$K41_{p,q} = k(s_{i_1+1}, t_{j_1+1}, x_{i_2}, y_{j_2}),$$
$$K42_{p,q} = k(s_{i_1+1}, t_{j_1+1}, x_{i_2}, y_{j_2+1}),$$
$$K43_{p,q} = k(s_{i_1+1}, t_{j_1+1}, x_{i_2+1}, y_{j_2}),$$
$$K44_{p,q} = k(s_{i_1+1}, t_{j_1+1}, x_{i_2+1}, y_{j_2+1}).$$

In this paper we supposed that $m_1 = m_2 = m_3 = m_4 = M$, for convergence.

3. THE BASIC CONCEPTS OF FUZZY EQUATIONS

We now recall some definitions needed through the paper.

DEFINITION 1. A fuzzy number is a set v:$RF \rightarrow I = [0,1]$ which satisfies in the following statements:

- v is upper semi continuous,
- $v(x) = 0$ outside of some interval such as $[c,d]$,
- There are real numbers a,b that $c \leq a \leq b \leq d$ and
- $v(x)$ is monotonic increasing on $[c,a]$,
- $v(x)$ is monotonic decreasing on $[b,d]$,
- $v(x) = 1$, $a \leq x \leq b$.

The set of all such fuzzy number is denoted by RF (Kaleva, 1987).

DEFINITION 2.

Let V be a fuzzy set on RF. V is called a fuzzy interval if:

- V is normal, there exists $x_0 \in RF$ such that $V(x_0) = 1$.
- V is convex, it holds that $V(\lambda x + (1 - \lambda)t) \geq min\{V(x),V(t)\}$, for all $x,t \in RF$ and $0 \leq \lambda \leq 1$.
- V is upper semi-continuous, $V(x_0) \geq \lim_{x \to x_0^{\pm}} V(x)$, for any $x_0 \in RF$,
- $[V]^0 = Cl\{x \in RF | V(x) > 0\}$ is a compact subset of RF.

The α-cut of a fuzzy interval V, with $0 < \alpha \leq 1$ is the crisp set, $[V]\alpha = \{x \in RF | V(x) > \alpha\}$.

For a fuzzy interval V, its α-cuts are closed intervals in RF (Kaleva, 1987). Let denote them by $[V]^{\alpha} = [\underline{V}(\alpha), \overline{V}(\alpha)]$.

An alternative definition or parametric form of a fuzzy number which yields the same *RF* is given by Kaleva (1987) as follows:

DEFINITION 3.

An arbitrary fuzzy number $\tilde{u}$ in the parametric form is represented by an ordered pair of functions $(\underline{u}(r), \overline{u}(r))$ which satisfy the following requirements (Ralescu and Adams (1980)):

- $\underline{u}(r)$ is a bounded left-continuous non-decreasing function over [0,1],
- $\overline{u}(r)$ is a bounded right-continuous non-increasing function over [0,1],
- $\underline{u}(r) \le \overline{u}(r)$, for all $0 \le r \le 1$.

For arbitrary fuzzy numbers $\tilde{v} = (\underline{v}(r), \overline{v}(r))$, $\tilde{w} = (\underline{w}(r), \overline{w}(r))$ and real number λ, we can define the addition and the scalar multiplication of the fuzzy numbers by using the extension principle as follows:

- $\tilde{v} = \tilde{w}$ if and only if $\underline{v}(r) = \underline{w}(r)$ and $\overline{v}(r) = \overline{w}(r)$,
- $\tilde{v} \oplus \tilde{w} = (\underline{v}(r) + \underline{w}(r), \overline{v}(r) + \overline{w}(r))$,

$$(\lambda \otimes \tilde{v}) = \begin{cases} (\lambda \underline{v}(r), \lambda \overline{v}(r)) & \lambda \ge 0, \\ (\lambda \overline{v}(r), \lambda \underline{v}(r)) & \lambda < 0. \end{cases}$$

DEFINITION 4.

For arbitrary numbers $\tilde{v} = (\underline{v}(r), \overline{v}(r))$ and $\tilde{w} = (\underline{w}(r), \overline{w}(r))$ (Nanda, 1989)

$$D(\tilde{v}, \tilde{w}) = max\left\{ \sup_{0\leq r\leq 1} |\overline{v}(r) - \overline{w}(r)|, \sup_{0\leq r\leq 1} |\underline{v}(r) - \underline{w}(r)| \right\},$$

in the distance between $\tilde{v}$ and $\tilde{w}$. It is proved that (*RF,D*) is a complete metric space with the properties (Nanda, 1989).

$$\forall \tilde{u}, \tilde{v}, \tilde{w} \in R_F \quad ; \quad D(\tilde{u} \oplus \tilde{w}, \tilde{v} \oplus \tilde{w}) = D(\tilde{u}, \tilde{v}),$$

$$\forall \tilde{u}, \tilde{v} \in R_F, \quad \forall k \in R \quad ; \quad D(k \otimes \tilde{u}, k \otimes \tilde{v}) = |k| D(\tilde{u}, \tilde{v}),$$

$$\forall \tilde{u}, \tilde{v}, \tilde{w}, \tilde{e} \in R_F \quad ; \quad D(\tilde{u} \oplus \tilde{v}, \tilde{w} \oplus \tilde{e}) \leq D(\tilde{u}, \tilde{w}) + D(\tilde{v}, \tilde{e}).$$

DEFINITION 5.

Let $\tilde{f}, \tilde{g} : [a, b] \rightarrow R_F$, be fuzzy real number valued functions. The uniform distance between $\tilde{f}, \tilde{g}$ is defined by Wang (1984):

$$D_U(\tilde{f}, \tilde{g}) = sup\left\{ D(\tilde{f}(x), \tilde{g}(x)) \,|\, x \in [a, b] \right\}. \quad (21)$$

In Bede and Gal (2005), the authors proved that if the fuzzy function, $\tilde{f}(t)$, is continuous in the metric *D*, its definite integral exists and also,

$$\underline{\int_a^b f(t, r)dt} = \int_a^b \underline{f}(t, r)dt,$$

DEFINITION 6.

A fuzzy real number valued function $\tilde{f}:[a,b]\rightarrow R_F$ is said to be continuous in $x_0 \in [a,b]$, if for each $\varepsilon > 0$ there is $\delta > 0$ such that $D(\tilde{f}(x),\tilde{f}(x_0)) < \varepsilon$, whenever $x \in [a,b]$ and $|x - x_0| < \delta$(Wu, 2000). We say that $\tilde{f}$ is fuzzy continuous on $[a,b]$ if $\tilde{f}$ is continuous at each$x_0 \in [a,b]$ and denote the space of all such functions by $CF([a,b])$.

DEFINITION 7.

Let $\tilde{f}:[a,b]\rightarrow R_F$ be a bounded function, then function $\omega_{[a,b]}(\tilde{f},.): R^{+}\cup\{0\}\rightarrow R^{+}$,

$$\omega_{[a,b]}(\tilde{f},\delta) = sup\left\{D(\tilde{f}(x),\tilde{f}(y))\middle| x,y\in\left[a,b\right], |x-y|\leq\delta\right\}, \qquad (22)$$

where R^{+} is the set of positive real numbers, is called the modulus of continuity of $\tilde{f}$ on $[a,b]$ (Wu, 2000).

DEFINITION 8.

Let $\tilde{f}:[a,b]\rightarrow R_F$, $\tilde{f}$ is fuzzy-Riemann integrable to $I \in RF$ if for any $\varepsilon > 0$, there exists $\delta > 0$ such that for any division $P = \{[u,v] ; \xi\}$ of $[a,b]$ with the norms $\Delta(p)<\delta$, we have (Wu, 2000):

$$D(\sum_{p}{}^{*}(v-u)\otimes\tilde{f}(\xi), I) < \varepsilon, \qquad (23)$$

where $\sum^{*}$ denotes the fuzzy summation. In this case it is denoted by

$$I = (FR)\int_a^b \tilde{f}(x)dx$$
.

LEMMA 1.

If $\tilde{f}, \tilde{g} : [a,b] \subseteq R \to R_F$ are fuzzy continuous functions, then the function *F*:[*a*,*b*]→*RF* by $F(x) = D(\tilde{f}(x), \tilde{g}(x))$ is continuous on [*a*,*b*] and by (Wu, 2000).

$$D\left(\int_a^b \tilde{f}(x)dx, \int_a^b \tilde{g}(x)dx\right) \leq \int_a^b D(\tilde{f}(x), \tilde{g}(x))dx.$$

DEFINITION 9.

A function $\tilde{f} : R^2 \to R_F$ is called a fuzzy function in two-dimensional space. $\tilde{f}$ is said to be continuous (Rivaz and Yousefi (2012)), if for arbitrary fixed $t_0 \in R^2$ and $\varepsilon > 0$ a $\delta > 0$ exists such that

$$\|t - t_0\| < \delta \Rightarrow D(\tilde{f}(t), \tilde{f}(t_0)) < \varepsilon \quad ; \quad t = (x, y), t_0 = (x_0, y_0).$$

DEFINITION 10.

Let $\tilde{f} : [a,b] \times [c,d] \to R_F$. For each partition $p = \{x_1, x_2, \ldots, x_m\}$ of [*a*,*b*] and $q = \{y_1, y_2, \ldots, y_n\}$ of [*c*,*d*] and for arbitrary $\xi : x_{i-1} \leq \xi_i \leq x_i$, $2 \leq i \leq m$ and for arbitrary $\eta_j : y_{j-1} \leq \eta_j \leq y_j$, $2 \leq j \leq n$, let

$$R_p = \sum_{i=2}^{m} \sum_{j=2}^{n} \tilde{f}(\xi_i, \eta_j)(x_i - x_{i-1})(y_j - y_{j-1}).$$

The definite integral of $\tilde{f}(x, y)$ over [*a*,*b*] × [*c*,*d*] is

$$\int_c^d \int_a^b \tilde{f}(x, y)dxdy = lim R_p,$$

$$\left(\max_{2\le i\le m}|x_i - x_{i-1}|, \max_{2\le j\le n}|y_j - y_{j-1}|\right) \to (0,0),$$

provided that this limit exists in metric D (Rivaz and Yousefi (2012)).

If the function $\tilde{f}(x,y)$ is continuous in the metric D, it's definite that integral exists (Goetschel and Voxman (1986)). Furthermore

$$\left(\underline{\int_c^d \int_a^b f(x,y,r)dxdy}\right) = \int_c^d \int_a^b \underline{f}(x,y,r)dxdy,$$

$$\left(\overline{\int_c^d \int_a^b f(x,y,r)dxdy}\right) = \int_c^d \int_a^b \overline{f}(x,y,r)dxdy.$$

4. TWO-DIMENSIONAL FUZZY FREDHOLM INTEGRAL EQUATION

Two-dimensional Fredholm integral equation of the second kind is defined as the following from (Xie and Lin (2009))

$$u(x,y) = f(x,y) + \lambda \int_a^b \int_c^d K(x,y,s,t)u(s,t)dsdt. \tag{24}$$

The linear two-dimensional fuzzy Fredholm integral equation of the second kind (2D-FFIE-2) is

$$\tilde{u}(x,y) = \tilde{f}(x,y) \oplus \lambda \otimes \int_a^b \int_c^d K(x,y,s,t) \otimes \tilde{u}(s,t)dsdt, \tag{25}$$

where $k(x,y,s,t)$ is an arbitrary kernel function over $S = [a,b] \times [c,d] \times [a,b] \times [c,d]$ and $\tilde{u}(x,y)$ and $\tilde{f}(x,y)$ are fuzzy real valued functions and $\tilde{u}(x,y)$ is unknown.

Now, we introduce parametric form of a 2D-FFIE-2 with respect to definition (3).

Let $(\underline{f}(x,y,r),\overline{f}(x,y,r))$ and $(\underline{u}(x,y,r),\overline{u}(x,y,r))$, $0 \le r \le 1$, be parametric form of. $\tilde{f}(x,y)$and $\tilde{u}(x,y)$, respectively. Then parametric from of 2D-FFIE-2 is as follows:

$$\underline{u}(x,y,r) = \underline{f}(x,y,r) + \lambda \int_a^b \int_c^d v_1(x,y,s,t,\underline{u}(s,t,r),\overline{u}(s,t,r))dsdt, \tag{26}$$

$$\overline{u}(x,y,r) = \overline{f}(x,y,r) + \lambda \int_a^b \int_c^d v_2(x,y,s,t,\underline{u}(s,t,r),\overline{u}(s,t,r))dsdt. \tag{27}$$

Where

$$v_1(x,y,s,t,\underline{u}(s,t,r),\overline{u}(s,t,r)) = \begin{cases} k(x,y,s,t)\underline{u}(s,t,r) & k(x,y,s,t) \ge 0, \\ k(x,y,s,t)\overline{u}(s,t,r) & k(x,y,s,t) < 0, \end{cases}$$

and

$$v_2(x,y,s,t,\underline{u}(s,t,r),\overline{u}(s,t,r)) = \begin{cases} k(x,y,s,t)\overline{u}(s,t,r) & k(x,y,s,t) \ge 0, \\ k(x,y,s,t)\underline{u}(s,t,r) & k(x,y,s,t) < 0, \end{cases}$$

for $0 \le r \le 1$. We can see that Eq.s (26) and (27) are systems of Fredholm integral equations of the second kind with three variables in crisp case.

5. SOLVING LINEAR TWO-DIMENSIONAL FUZZY FREDHOLM INTEGRAL EQUATION BY 2D-TFS

In this section, we present an effective method for solving a 2D-FFIE-2 by using 2D-TFs. The parametric form of (25), with $a = c = 0$, $b = d = 1$ and $\lambda = 1$, is:

$$\underline{u}(x,y,r) = \underline{f}(x,y,r) + \int_0^1 \int_0^1 k(x,y,s,t)\underline{u}(s,t,r)dsdt, \tag{28}$$

$$\overline{u}(x,y,r) = \overline{f}(x,y,r) + \int_0^1 \int_0^1 k(x,y,s,t)\overline{u}(s,t,r)dsdt. \tag{29}$$

2D-TFs are applied for Eq.s (29) and (28) respectively. To explain this method for Eq.(28), first let us expand $\underline{u}(x,y,r)$, $\underline{f}(x,y,r)$ and $k(x,y,s,t)$ by 2D-TFs as follows:

$$\underline{u}(x,y,r) \approx T^T(x,y)UT(r), \tag{30}$$

$$\underline{f}(x,y,r) \approx T^T(x,y)FT(r), \tag{31}$$

$$k(x,y,s,t) \approx T^T(x,y)KT(s,t). \tag{32}$$

Where $T(x,y)$ and $T(r)$ are defined in Eq.s (9) and (15), respectively and U and F are $(4M^2 \times 2M)$-matrices of 2D-TFs coefficients of $\underline{u}(x,y,r)$ and $\underline{f}(x,y,r)$, respectively, and$K$ is $(4M^2 \times 4M^2)$-matrix 2D-TFs coefficients of $k(x,y,s,t)$.

To approximate the solution of Eq. (28), from Eq.s (32), (31) and (30), we have

$$T^T(x,y)UT(r) \approx T^T(x,y)FT(r) + \int_0^1 \int_0^1 T^T(x,y)KT(s,t)T^T(s,t)UT(r)dsdt.$$

$$T^T(x,y)UT(r) \approx T^T(x,y)FT(r) + T^T(x,y)K\left(\int_0^1\int_0^1 T(s,t)T^T(s,t)dsdt\right)UT(r),$$

using the Eq. (16) we have

$$T^T(x,y)UT(r) \approx T^T(x,y)FT(r) + T^T(x,y)KDUT(r).$$

then

$$U = F + KDU,$$

where U and F are $(4M^2 \times 2M)$-matrix and KD is $(4M^2 \times 4M^2)$-matrix, so KDU is $(4M^2 \times 2M)$-matrix, where U is unknown. Then we have

$$(I - KD)U = F,$$

where $I_{4M^2 \times 4M^2}$ is identity matrix. Then we can write above equation as follows

$$U = (I - KD)^{-1}F.$$

By solving this matrix system we can find matrix $U_{4M^2 \times 2M}$ and we can approximate $u(x,y,r)$ from

$$u(x,y,r) = T^T(x,y)UT(r).$$

The same trend holds for Eq. (29).

6. THE CONVERGENCE OF THE METHOD

In this section, we obtain error estimate for the numerical method proposed in previous section.

THEOREM 1.

The solution of two-dimensional Fredholm fuzzy integral equations such as Eq. (25), by using 2D-TFs are convergence to exact solution if

$$P = \max_{0 \le x,y,s,t \le 1} |k(x,y,s,t)| < 1.$$

PROOF. Assume $\tilde{u}(x,y)$ and $\tilde{u}_M(x,y)$ show approximate and exact solution of Eq. (25)respectively. Then

$$\begin{aligned} D(\tilde{u}(x,y),\hat{u}_M(x,y)) &= D\Big(\int_0^1\int_0^1 k(x,y,s,t)\tilde{u}(s,t)dsdt, \int_0^1\int_0^1 k(x,y,s,t)\Big(\sum_{i=0}^{M-1}{}^*\sum_{j=0}^{M-1}{}^* c_{i,j}T_{i,j}^{1,1}(s,t) + \sum_{i=0}^{M-1}{}^*\sum_{j=0}^{M-1}{}^* d_{i,j}T_{i,j}^{1,2}(s,t) + \sum_{i=0}^{M-1}{}^*\sum_{j=0}^{M-1}{}^* e_{i,j}T_{i,j}^{2,1}(s,t) \\ &+ \sum_{i=0}^{M-1}{}^*\sum_{j=0}^{M-1}{}^* l_{i,j}T_{i,j}^{2,2}(s,t)\Big)dsdt\Big) \le \int_0^1\int_0^1 D\Big(k(x,y,s,t)\tilde{u}(s,t), k(x,y,s,t)\Big(\sum_{i=0}^{M-1}{}^*\sum_{j=0}^{M-1}{}^* c_{i,j}T_{i,j}^{1,1}(s,t) + \sum_{i=0}^{M-1}{}^*\sum_{j=0}^{M-1}{}^* d_{i,j}T_{i,j}^{1,2}(s,t) \\ &+ \sum_{i=0}^{M-1}{}^*\sum_{j=0}^{M-1}{}^* e_{i,j}T_{i,j}^{2,1}(s,t) + \sum_{i=0}^{M-1}{}^*\sum_{j=0}^{M-1}{}^* l_{i,j}T_{i,j}^{2,2}(s,t)\Big)\Big)dsdt \le P\int_0^1\int_0^1 D\Big(\tilde{u}(s,t), \sum_{i=0}^{M-1}{}^*\sum_{j=0}^{M-1}{}^* c_{i,j}T_{i,j}^{1,1}(s,t) + \sum_{i=0}^{M-1}{}^*\sum_{j=0}^{M-1}{}^* d_{i,j}T_{i,j}^{1,2}(s,t) \\ &+ \sum_{i=0}^{M-1}{}^*\sum_{j=0}^{M-1}{}^* e_{i,j}T_{i,j}^{2,1}(s,t) + \sum_{i=0}^{M-1}{}^*\sum_{j=0}^{M-1}{}^* l_{i,j}T_{i,j}^{2,2}(s,t)\Big)dsdt = P\int_0^1\int_0^1 D(\tilde{u}(s,t),\tilde{u}_M(s,t))dsdt, \end{aligned}$$

therefore we have

$$D(\tilde{u}(x,y),\tilde{u}_M(x,y)) \le PD(\tilde{u}(x,y),\tilde{u}_M(x,y)),$$

$$\sup_{(x,y)\in\Omega} [0,1)\times[0,1))D(\tilde{u}(x,y),\tilde{u}_M(x,y)) \le P \sup_{(x,y)\in\Omega} [0,1)\times[0,1))D(\tilde{u}(x,y),\tilde{u}_M(x,y)).$$

Therefore, if $P < 1$, we will have:

$$\lim_{M\to\infty} \sup_{(x,y)\in\Omega} D(\tilde{u}(x,y),\tilde{u}_M(x,y)) = 0.$$

7. NUMERICAL EXAMPLES

Here, two examples are given to certify the convergence and error bound of the 2D-TFs method for Two-dimensional fuzzy Fredholm integral equations. All results are computed by using a program written in the Matlab. In this regard, we have presented with tables and figures.

EXAMPLE 1.

Consider the following 2D-FFIE-2 (Rivaz and Yousefi (2012)).

$$\underline{f}(x,y,r) = x\sin(\frac{y}{2})(r^2+r),$$

$$\overline{f}(x,y,r) = x\sin(\frac{y}{2})(4-r^3-r),$$

and

$$k(x,y,s,t) = x^2ys, \quad 0 \le x,y,s,t \le 1 \quad and\ \lambda = 1.$$

That the exact solution is

$$\underline{u}(x,y,r) = (x\sin(\frac{y}{2}) - \frac{16}{21}(\cos(\frac{1}{2})-1)x^2y)(r^2+r),$$

$$\overline{u}(x,y,r) = (x\sin(\frac{y}{2}) - \frac{16}{21}(\cos(\frac{1}{2})-1)x^2y)(4-r^3-r).$$

Table 1 and Table 2 show the exact solution and the numerical solutions by the presented method and absolute error of this method. Fig. 1 and Fig. 2 show comparison between the exact solution and the numerical solutions by the presented method.

EXAMPLE 2.

Consider the following 2D-FFIE-2

$$\underline{f}(x,y,r) = r(xy+\frac{1}{676}(x^2+y^2-2)),$$

$$\overline{f}(x,y,r) = (2-r)(xy+\frac{1}{676}(x^2+y^2-2)),$$

Table 1. Numerical results of Example 1 with 2D-TFs method.

r	Exact solution $(\underline{u}(x,y,r),\overline{u}(x,y,r))$	2D-TFs method for $x = 0.1$, $y = 0.1$ and $M = 5$	2D-TFs method for $x = 0.1$, $y = 0.1$ and $M = 10$	Absolute error for $x = 0.1$, $y = 0.1$ and $M = 10$
0.0	(0.00000000, 0.02036474)	(0.00000000, 0.02071364)	(0.00000000, 0.02036484)	(0.00000000e-00,9.9976971e-08)
0.1	(0.00056003, 0.01985053)	(0.00062140, 0.02017509)	(0.00056003, 0.01985063)	(2.7493667e-09, 9.7452553e-08)
0.2	(0.00122188, 0.01930578)	(0.00124281, 0.01963653)	(0.00122189, 0.01930587)	(5.9986183e-09, 9.4778169e-08)
0.3	(0.00198556, 0.01869993)	(0.00207136, 0.01897370)	(0.00198557, 0.01870002)	(9.7477547e-09, 9.1803854e-08)
0.4	(0.00285106, 0.01800243)	(0.00289991, 0.01831086)	(0.00285107, 0.01800252)	(1.3996776e-08, 8.8379643e-08)
0.5	(0.00381839, 0.01718275)	(0.00393559, 0.01739946)	(0.00381840, 0.01718284)	(1.8745682e-08, 8.4355570e-08)
0.6	(0.00488753, 0.01621034)	(0.00497127, 0.01648806)	(0.00488756, 0.01621042)	(2.3994473e-08, 7.9581669e-08)
0.7	(0.00605851, 0.01505464)	(0.00621409, 0.01520381)	(0.00605854, 0.01505471)	(2.9743149e-08, 7.3907976e-08)
0.8	(0.00733130, 0.01368511)	(0.00745691, 0.01391957)	(0.00733134, 0.01368517)	(3.5991709e-08, 6.7184525e-08)
0.9	(0.00870593, 0.01207120)	(0.00890686, 0.01213819)	(0.00870597, 0.01207126)	(4.2740155e-08, 5.9261350e-08)

Table 2. Numerical results of Example 1 with 2D-TFs method.

r	Exact solution $(\underline{u}(x,y,r),\overline{u}(x,y,r))$	2D-TFs method for $x = 0.3$, $y = 0.6$ and $M = 10$	Absolute error for $x = 0.3$, $y = 0.6$ and $M = 10$	Absolute error the method of Rivaz and Yousefi (2012) for $n = 52$
0.0	(0.00000000, 0.37477066)	(0.00000000, 0.37477605)	(0.0000000e-00, 5 3987564e-06)	(0.0009, 0.0000)
0.1	(0.01030619, 0.36530770)	(0.01030634, 0.36531296)	(1.4846580e-07, 5 2624378e-06)	(0.0007, 0.0004)
0.2	(0.02248623, 0.35528258)	(0.02248656, 0.35528770)	(3.2392538e-07, 5 1180211e-06)	(0.0003, 0.0007)
0.3	(0.03654013, 0.34413315)	(0.03654066, 0.34413811)	(5.2637875e-07, 4 9574081e-06)	(0.0002, 0.0011)
0.4	(0.05246789, 0.33129726)	(0.05246864, 0.33130203)	(7.5582590e-07, 4 7725007e-06)	(0.0009, 0.0016)
0.5	(0.07026949, 0.31621274)	(0.07027051, 0.31621274)	(1.0122668e-06, 4 5552007e-06)	(0.0020, 0.0008)
0.6	(0.08994495, 0.29831744)	(0.08994625, 0.29832174)	(1.2957015e-06, 4 2974101e-06)	(0.0015, 0.0004)
0.7	(0.11149427, 0.27704921)	(0.11149587, 0.27705320)	(1.6061300e-06, 3 9910307e-06)	(0.0007, 0.0003)
0.8	(0.13491743, 0.25184588)	(0.13491938, 0.25184951)	(1.9435523e-06, 3 6279643e-06)	(0.0001, 0.0012)
0.9	(0.16021445, 0.22214530)	(0.16021676, 0.22214850)	(2.3079683e-06, 3 2001129e-06)	(0.0011, 0.0024)

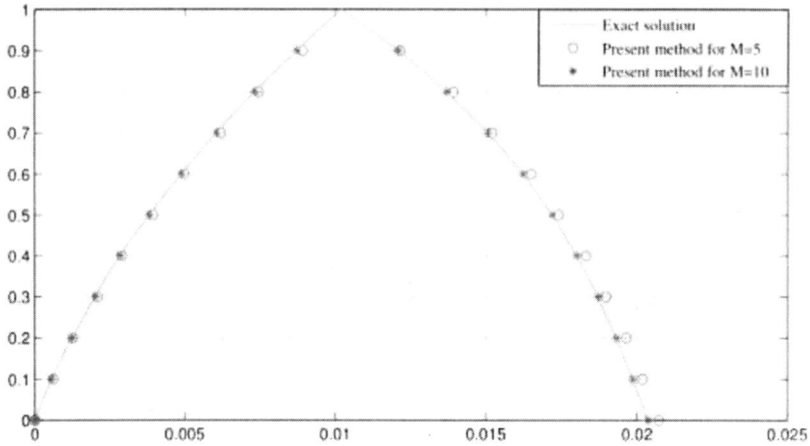

Figure 1. Comparison between the exact solution and the approximate solution by present method for M = 5, 10.

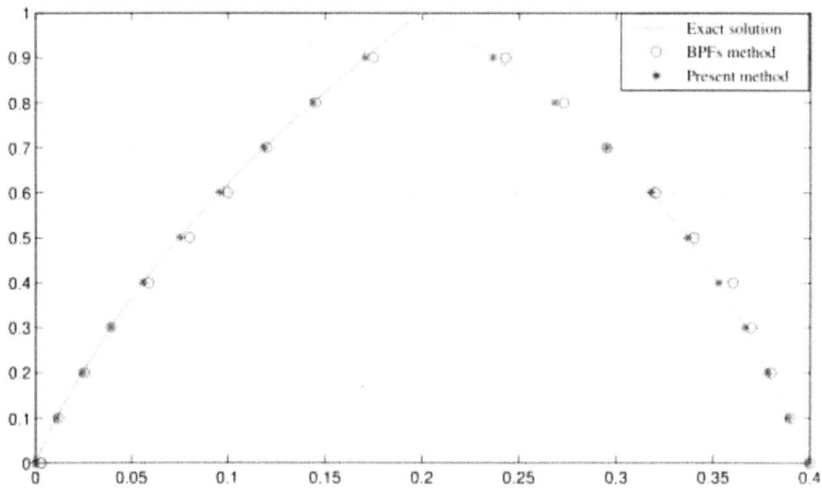

Figure 2. Comparison between the exact solution, the approximate solution by BPFs method for n = 52 and present method for M = 10.

and

$$k(x,y,s,t)=\frac{1}{169}(x^2+y^2-2)(s^2+t^2-2),\quad 0\le x,y,s,t\le 1\quad and\,\lambda=1.$$

The exact solution is

$$\underline{u}(x,y,r)=rxy,$$

$$\overline{u}(x,y,r)=(2-r)xy.$$

Absolute values of the error function are provided in Table 3 by taking $M = 4,\ 12$. The numerical histories for $N = 12$ is depicted in Fig. 3.

8. CONCLUSION

Fuzzy integral equations are important for studying and solving a large proportion of the problems in many topics in applied mathematics, particularly in fuzzy control. In this study, we introduce 2D-TFs method for approximating the solution of linear (2D-FFIE-2). By this method, the original equation is converted into two crisp (2D-FFIE-2). The efficiency and simplicity of this method are illustrated by introducing some numerical examples with known exact solutions. The main advantage of this method is low cost of setting up the equations without using any projection method and any integration.

Table 3. Numerical results of Example 2 with 2D-TFs method method.

r	Exact solution $(\underline{u}(x,y,r), \overline{u}(x,y,r))$	2D-TFs method for $x = 0.1$, $y = 0.4$ and $M = 4$	2D-TFs method for $x = 0.1$, $y = 0.4$ and $M = 12$	Absolute error for $x = 0.1$, $y = 0.4$ and $M = 12$
0.0	(0.00000000, 0.08000000)	(0.00000000, 0.07973560)	(0.00000000, 0.08000985)	(0.0000000e-00, 9.8553102e-06)
0.1	(0.00400000, 0.07600000)	(0.00398678, 0.07574882)	(0.00400049, 0.07600936)	(4.9276551e-07, 9.3625447e-06)
0.2	(0.00800000, 0.07200000)	(0.00797356, 0.07176204)	(0.00800098, 0.07200886)	(9.8553102e-07, 8.8697792e-06)
0.3	(0.01200000, 0.06800000)	(0.01196034, 0.06777526)	(0.01200147, 0.06800837)	(1.4782965e-06, 8.3770137e-06)
0.4	(0.01600000, 0.06400000)	(0.01594712, 0.06378849)	(0.01600197, 0.06400788)	(1.9710620e-06, 7.8842481e-06)
0.5	(0.02000000, 0.06000000)	(0.01993390, 0.05980170)	(0.02000246, 0.06000739)	(2.4638275e-06, 7.3914826e-06)
0.6	(0.02400000, 0.05600000)	(0.02392068, 0.05581492)	(0.02400295, 0.05600689)	(2.9565930e-06, 6.8987171e-06)
0.7	(0.02800000, 0.05200000)	(0.02790746, 0.05182814)	(0.02800344, 0.05200640)	(3.4493585e-06, 6.4059516e-06)
0.8	(0.03200000, 0.04800000)	(0.03189424, 0.04784136)	(0.0320039, 0.04800591)	(3.9421240e-06, 5.9131861e-06)
0.9	(0.03600000, 0.04400000)	(0.03588102, 0.04385458)	(0.03600443, 0.04400542)	(4.4348896e-06, 5.4204206e-06)

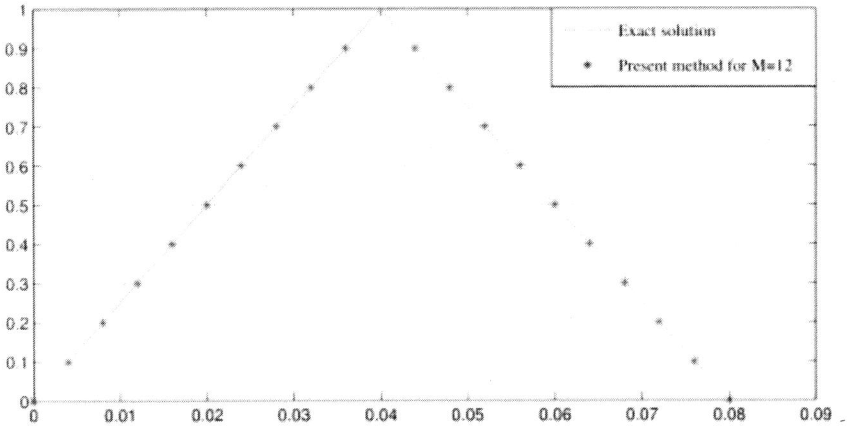

Figure. 3. Comparison between the exact solution and the approximate solution by present method for $M = 12$.

ACKNOWLEDGMENT

The authors are very grateful to the editor and the referees for their helpful comments and valuable suggestions.

REFERENCES

1. Babolian E, Mokhtari R, Salmani M. Using direct method for solving variational problems via triangular orthogonal functions. Appl Math Comput 2007;191:206e17.
2. Babolian E, Masouri Z, Hatamzadeh-Varmazyar S. Numerical solution of nonlinear Volterra-Fredholm integro-differential equations via direct method using triangular functions. Comput Math Appl 2009;58:239e47.
3. Babolian E, Maleknejad K, Roodaki M, Almasieh H. Two-dimensional triangular functions and their applications to

nonlinear 2D Volterra-Fredholm integral equations. Comput Math Appl 2010;60:1711e22.

4. Bede B, Gal SG. Generalizations of the diferentiability of fuzzy number valued function with applications to fuzzy differential equations. Fuzzy Sets Syst 2005;151:581e99.
5. Deb A, Dasgupta A, Sarkar G. A new set of orthogonal functions and its applications to the analysis of dynamic systems. J Frankl Inst 2006;343:1e26.
6. Ezzati R, Ziari S. Numerical solution of two-dimensional fuzzy fredholm integral equation of the second kind using fuzzy brivariate Bernstein polynomials. Int J Fuzzy Syst 2013;15:84e8.
7. Goetschel R, Voxman W. Elementary fuzzy calculus. Fuzzy Sets Syst 1986;18:31e43.
8. Kaleva O. Fuzzy differential equations. Fuzzy Sets Syst 1987;24:301e17.
9. Mirzaee F, Bimesl S. Application of Euler matrix method fo solving linear and a class of nonlinear Fredholm integro-differential equations. Mediterr J Math 2014;11(3):999e1018.
10. Molabahrami A, Shidfar A, Ghyasi A. An analytical method for solving linear Fredholm fuzzy integral equations of the second kind. Comput Math Appl 2011;61:2754e61.
11. Nanda S. On integration of fuzzy mappings. Fuzzy Sets Syst 1989;32:95e101
12. Ralescu D, Adams G. The fuzzy integrals. J Math Anal Appl 1980;75:562e70.
13. Rivaz A, Yousefi F. Modified homotopy perturbation method for solving two-dimensional fuzzy Fredholm integral equation. Int J Appl Math 2012;25:591e602.
14. Seikkala S. On the fuzzy initial value problem. Fuzzy Sets Syst 1987;24:319e30.
15. Wang Z. The autocontinuity of set-function and the fuzzy integral. J Math Analysis Appl 1984;99:195e218.

16. Wu Hc. The fuzzy Riemann integral and numerical integration.
17. Fuzzy Sets Syst 2000;110:1e25.
18. Xie W, Lin F. A fast numerical solution method for two-dimensional Fredholm integral equations of the second kind. Appl Numer Math 2009;59:1709e19.

CHAPTER 6

Inequalities for modified Bessel functions and their integrals

Robert E. Gaunt[1,]

Department of Statistics, University of Oxford, Oxford, UK

ABSTRACT

Simple inequalities for some integrals involving the modified Bessel functions $I_\nu(x)$ and $K_\nu(x)$ are established. We also obtain a monotonicity result for $K_\nu(x)$ and a new lower bound, that involves gamma functions, for $K_0(x)$.

KEYWORDS

Modified Bessel functions

1. INTRODUCTION AND PRELIMINARY RESULTS

In the developing Stein's method for Variance-Gamma distributions, Gaunt [5] required simple bounds, in terms of modified Bessel functions, for the integrals

$$\int_0^x e^{\beta t} t^\nu I_\nu(t)\,dt \quad \text{and} \quad \int_x^\infty e^{\beta t} t^\nu K_\nu(t)\,dt,$$

where $x>0$, $\nu>-1/2$ and $-1<\beta<1$. Closed form expressions for these integrals, in terms of modified Bessel functions and the modified Struve function $\mathbf{L}_\nu(x)$, do in fact exist for the case $\beta=0$. For $z\in C$ and $\nu\in C$, let $L_\nu(z)$ denote $I_\nu(z)$, $e^{\nu\pi i}K_\nu(z)$ or any linear combination of these functions, in which the coefficients are independent of ν and z. From formula 10.43.2 of Olver et al. [13] we have, for $\nu\neq-1/2$,

$$\int z^\nu \mathscr{L}_\nu(z)\,dz = \sqrt{\pi}2^{\nu-1}\Gamma(\nu+1/2)z\left(\mathscr{L}_\nu(z)\mathbf{L}_{\nu-1}(z) - \mathscr{L}_{\nu-1}(z)\mathbf{L}_\nu(z)\right). \tag{1.1}$$

Whilst formula (1.1) holds for complex-valued z and ν, throughout this paper we shall restrict our attention to the case of real-valued z and ν . There are no closed form expressions in terms of modified Bessel and Struve functions in the literature for the integrals for the case $\beta\neq0$. Moreover, even in the case $\beta=0$ the expression on the right-hand side of formula (1.1) is a complicated expression involving the modified Struve function $\mathbf{L}_\nu(x)$. This provides the motivation for establishing simple bounds, in terms of modified Bessel functions, for the integrals defined in the first display.

In this paper we establish, through the use of elementary properties of modified Bessel functions and straightforward calculations, simple bounds, that involve modified Bessel functions, for the integrals given in the first display. Our bounds prove to be very useful when applied to calculations that arise in the study of Stein's method for Variance-Gamma distributions. We also obtain a monotonicity result and bound for the modified Bessel function of the second kind $K_\nu(x)$, as well as a simple but remarkably tight lower bound for $K_0(x)$. These bounds are, again, motivated by the need for such bounds in the study of Stein's method for Variance-Gamma distributions. However, the bounds obtained in this paper may also prove to be useful in other problems related to modified Bessel functions; see, for example, Baricz and Sun [4] in which inequalities for modified Bessel functions of the first kind were used to obtain lower and upper bounds for integrals of involving modified Bessel functions of the first kind. Throughout this paper we make use of some elementary properties of modified Bessel functions and these are stated inAppendix A.

2. INEQUALITIES FOR INTEGRALS INVOLVING MODIFIED BESSEL FUNCTIONS

Before presenting our first result concerning inequalities for integrals of modified Bessel functions, we introduce some notation for the repeated integral of the function$e^{\beta x}x^\nu I_\nu(x)$, which will be used in the following theorem. We define

$$I_{(\nu,\beta,0)}(x) = e^{\beta x}x^\nu I_\nu(x), \qquad I_{(\nu,\beta,n+1)}(x) = \int_0^x I_{(\nu,\beta,n)}(y)\,dy, \quad n = 0,1,2,3,\ldots. \tag{2.1}$$

With this notation we have:

THEOREM 2.1.

Let $0 \leq \gamma < 1$, *then the following inequalities hold for all* $x > 0$

$$\int_0^x t^\nu I_\nu(t)\,\mathrm{d}t > x^\nu I_{\nu+1}(x), \quad \nu > -1, \tag{2.2}$$

$$\int_0^x t^\nu I_\nu(t)\,\mathrm{d}t < x^\nu I_\nu(x), \quad \nu \geq 1/2, \tag{2.3}$$

$$I_{(\nu,0,n+1)}(x) < I_{(\nu,0,n)}(x), \quad \nu \geq 1/2, \tag{2.4}$$

$$I_{(\nu,-\gamma,n)}(x) \leq \frac{1}{(1-\gamma)^n} \mathrm{e}^{-\gamma x} I_{(\nu,0,n)}(x), \quad \nu \geq 1/2, \ n = 0, 1, 2, \ldots, \tag{2.5}$$

$$\int_0^x t^\nu I_{\nu+n}(t)\,\mathrm{d}t < \frac{2(\nu+n+1)}{2\nu+n+1} x^\nu I_{\nu+n+1}(x), \quad \nu > -1/2, \ n \geq 0, \tag{2.6}$$

$$I_{(\nu,0,n)}(x) < \left\{ \prod_{k=1}^{n} \frac{2\nu+2k}{2\nu+k} \right\} x^\nu I_{\nu+n}(x), \quad \nu \geq 0, \ n = 1, 2, 3, \ldots, \tag{2.7}$$

$$I_{(\nu,-\gamma,n)}(x) < \frac{1}{(1-\gamma)^n} \left\{ \prod_{k=1}^{n} \frac{2\nu+2k}{2\nu+k} \right\} \mathrm{e}^{-\gamma x} x^\nu I_{\nu+n}(x), \quad \nu \geq 1/2, \ n = 1, 2, 3, \ldots.$$

PROOF.

(i) From the differentiation formula (A.12) we have that

$$\int_0^x t^\nu I_\nu(t)\,\mathrm{d}t = \int_0^x \frac{1}{t} t^{\nu+1} I_\nu(t)\,\mathrm{d}t > \frac{1}{x}\int_0^x t^{\nu+1} I_\nu(t)\,\mathrm{d}t = x^\nu I_{\nu+1}(x),$$

since by (A.2) we have $\lim_{x\downarrow 0} x^{\nu+1} I_{\nu+1}(x) = 0$ for $\nu > -1$.

(ii) Using inequality (A.7) and then applying (A.12) we get

$$\int_0^x t^\nu I_\nu(t)\,\mathrm{d}t < \int_0^x t^\nu I_{\nu-1}(t)\,\mathrm{d}t = x^\nu I_\nu(x).$$

(iii) From inequality (2.3), we have

$$I_{(\nu,0,1)}(x) < I_{(\nu,0,0)}(x).$$

Integrating both sides of the above display *n* times with respect to *x* yields the desired inequality.

(iv) We prove the result by induction on *n* . The result is trivially true for $n=0$. Suppose the result is true for $n=k$. From the inductive hypothesis we have

$$I_{(\nu,-\gamma,k+1)}(x) = \int_0^x I_{(\nu,-\gamma,k)}(t)\,\mathrm{d}t \le \frac{1}{(1-\gamma)^k}\int_0^x \mathrm{e}^{-\gamma t} I_{(\nu,0,k)}(t)\,\mathrm{d}t. \tag{2.8}$$

Integration by parts and inequality (2.4) gives

$$\int_0^x \mathrm{e}^{-\gamma t} I_{(\nu,0,k)}(t)\,\mathrm{d}t = \mathrm{e}^{-\gamma x} I_{(\nu,0,k+1)}(x) + \gamma\int_0^x \mathrm{e}^{-\gamma t} I_{(\nu,0,k+1)}(t)\,\mathrm{d}t$$

$$< \mathrm{e}^{-\gamma x} I_{(\nu,0,k+1)}(x) + \gamma\int_0^x \mathrm{e}^{-\gamma t} I_{(\nu,0,k)}(t)\,\mathrm{d}t.$$

Rearranging we obtain

$$\int_0^x e^{-\gamma t} I_{(\nu,0,k)}(t)\,dt < \frac{1}{1-\gamma} e^{-\gamma x} I_{(\nu,0,k+1)}(x),$$

and substituting into (2.8) gives

$$I_{(\nu,-\gamma,k+1)}(x) < \frac{1}{(1-\gamma)^{k+1}} e^{-\gamma x} I_{(\nu,0,k+1)}(x).$$

Hence the result has been proved by induction.

(v) From the differentiation formula (A.12) and identity (A.10) we get that

$$\begin{aligned}\frac{d}{dt}\left(t^{\nu} I_{\nu+n+1}(t)\right) &= \frac{d}{dt}\left(t^{-(n+1)} \cdot t^{\nu+n+1} I_{\nu+n+1}(t)\right)\\ &= t^{\nu} I_{\nu+n}(t) - (n+1)t^{\nu-1} I_{\nu+n+1}(t)\\ &= t^{\nu} I_{\nu+n}(t) - \frac{n+1}{2(\nu+n+1)} t^{\nu} I_{\nu+n}(t) + \frac{n+1}{2(\nu+n+1)} t^{\nu} I_{\nu+n+2}(t)\\ &= \frac{2\nu+n+1}{2(\nu+n+1)} t^{\nu} I_{\nu+n}(t) + \frac{n+1}{2(\nu+n+1)} t^{\nu} I_{\nu+n+2}(t).\end{aligned}$$

Integrating both sides over $(0,x)$, applying the fundamental theorem of calculus and rearranging give

$$\int_0^x t^{\nu} I_{\nu+n}(t)\,dt = \frac{2(\nu+n+1)}{2\nu+n+1} x^{\nu} I_{\nu+n+1}(x) - \frac{n+1}{2\nu+n+1}\int_0^x t^{\nu} I_{\nu+n+2}(t)\,dt.$$

The result now follows from the fact that $I_\nu(x)>0$ for $x>0$ and by the positivity of the integral.

(vi) From inequality (2.6) we have

$$I_{(\nu,0,1)}(x) = \int_0^x t^{\nu} I_{\nu}(t)\,dt < \frac{2(\nu+1)}{2\nu+1} x^{\nu} I_{\nu+1}(x),$$

and

$$
\begin{aligned}
I_{(\nu,0,2)}(x) &= \int_0^x I_{(\nu,0,1)}(t)\,\mathrm{d}t \\
&< \frac{2(\nu+1)}{2\nu+1}\int_0^x t^\nu I_{\nu+1}(t)\,\mathrm{d}t \\
&< \frac{2(\nu+1)}{2\nu+1}\frac{2(\nu+2)}{2\nu+2}x^\nu I_{\nu+2}(x).
\end{aligned}
$$

Iterating gives the result.

(vii) This follows from inequalities (2.5) and (2.7).

We now state a simple lemma (which is a special case of Lemma 2.4 of Ismail and Muldoon [9]), that gives a monotonicity result for the ratio $\frac{K_{\nu-1}(x)}{K_\nu(x)}$. The lemma has an immediate corollary, which we will make use of in the proof of our next theorem.

LEMMA 2.2.

Suppose $x>0$, *then the function* $\frac{K_{\nu-1}(x)}{K_\nu(x)}$ *is strictly monotone increasing for* $\nu>1/2$, *is constant for* $\nu=1/2$, *and is strictly monotone decreasing for* $\nu<1/2$.

COROLLARY 2.3.

For $\nu>1/2$ *and* $\alpha>1$ *the equation* $K_\nu(x)=\alpha K_{\nu-1}(x)$ *has one root in the region* $x>0$.

PROOF.

From the asymptotic formulas (A.3) and (A.5), it follows that for $\nu > 1/2$,

$$\lim_{x \downarrow 0} \frac{K_{\nu-1}(x)}{K_\nu(x)} = 0, \quad \text{and} \quad \lim_{x \to \infty} \frac{K_{\nu-1}(x)}{K_\nu(x)} = 1.$$

Since $\frac{K_{\nu-1}(x)}{K_\nu(x)}$ is strictly monotone increasing on $(0,\infty)$, it follows that for $\alpha > 1$ the equation $K_\nu(x) = \alpha K_{\nu-1}(x)$ (i.e. $\frac{K_{\nu-1}(x)}{K_\nu(x)} = \frac{1}{\alpha}$) has one root in the region $x > 0$.

As an aside, we note that Lemma 2.3 allows us to easily establish an inequality for the Turánian $\Delta_\nu(x) = K_\nu^2(x) - K_{\nu-1}(x)K_{\nu+1}(x)$ (for more details on the Turánian $\Delta_\nu(x)$ see Baricz [1]).

PROPOSITION 2.4.

Suppose $x > 0$, *then* $\Delta_\nu(x) < \Delta_{\nu-1}(x)$ *for* $\nu > 1/2$, $\Delta_{1/2}(x) = \Delta_{-1/2}(x)$, *and* $\Delta_\nu(x) > \Delta_{\nu-1}(x)$ *for* $\nu < 1/2$.

PROOF.

By the quotient rule and differentiation formula (A.14), we have

$$\begin{aligned}
\frac{\mathrm{d}}{\mathrm{d}x}\left(\frac{K_{\nu-1}(x)}{K_\nu(x)}\right) &= -\frac{(K_\nu(x) + K_{\nu-2}(x))K_\nu(x) - (K_{\nu+1}(x) + K_{\nu-1}(x))K_{\nu-1}(x)}{2K_\nu^2(x)} \\
&= \frac{K_{\nu-1}^2(x) - K_{\nu-2}(x)K_\nu(x) - (K_\nu^2(x) - K_{\nu-1}(x)K_{\nu+1}(x))}{2K_\nu^2(x)} \\
&= \frac{\Delta_{\nu-1}(x) - \Delta_\nu(x)}{2K_\nu^2(x)}.
\end{aligned}$$

Since, by Lemma 2.3, the function $\frac{K_{\nu-1}(x)}{K_\nu(x)}$ is strictly monotone increasing for $\nu>1/2$, is constant for $\nu=1/2$, and is strictly monotone decreasing for $\nu<1/2$, the result follows. □

With the aid of Corollary 2.3 and standard properties of the modified Bessel function $K_\nu(x)$, we can prove at the following theorem.

THEOREM 2.5. *Let* $-1<\beta<1$, *then for all* $x>0$ *the following inequalities hold*

$$\int_x^\infty t^\nu K_\nu(t)\,\mathrm{d}t < x^\nu K_{\nu+1}(x), \quad \nu \in \mathbb{R},$$

$$\int_x^\infty t^\nu K_\nu(t)\,\mathrm{d}t < x^\nu K_\nu(x), \quad \nu < 1/2, \tag{2.9}$$

$$\int_x^\infty e^{\beta t} t^\nu K_\nu(t)\,\mathrm{d}t < \frac{1}{1-|\beta|} e^{\beta x} x^\nu K_\nu(x), \quad \nu < 1/2, \tag{2.10}$$

$$\int_x^\infty t^\nu K_\nu(t)\,\mathrm{d}t \le \frac{\sqrt{\pi}\,\Gamma(\nu+1/2)}{\Gamma(\nu)} x^\nu K_\nu(x), \quad \nu \ge 1/2, \tag{2.11}$$

$$\int_x^\infty e^{\beta t} t^\nu K_\nu(t)\,\mathrm{d}t \le \frac{2\sqrt{\pi}\,\Gamma(\nu+1/2)}{(1-\beta^2)^{\nu+1/2}\Gamma(\nu)} e^{\beta x} x^\nu K_\nu(x), \quad \nu \ge 1/2.$$

PROOF.

(i) From the differentiation formula (A.13) we have that

$$\int_x^\infty t^\nu K_\nu(t)\,\mathrm{d}t = \int_x^\infty \frac{1}{t} t^{\nu+1} K_\nu(t)\,\mathrm{d}t < \frac{1}{x}\int_x^\infty t^{\nu+1} K_\nu(t)\,\mathrm{d}t = x^\nu K_{\nu+1}(x),$$

since, by the asymptotic formula (A.5), $\lim_{x\to\infty} x^{\nu+1}K_{\nu+1}(x) = 0$.

(ii) Using inequality (A.8) and then applying the differentiation formula (A.13) we have

$$\int_x^\infty t^\nu K_\nu(t)\,\mathrm{d}t < \int_x^\infty t^\nu K_{\nu-1}(t)\,\mathrm{d}t = x^\nu K_\nu(x).$$

(iii) Now suppose that $\nu<1/2$ and $\beta>0$. Using integration by parts and the differentiation formula (A.13) gives

$$\int_x^\infty e^{\beta t} t^\nu K_\nu(t)\,\mathrm{d}t = -\frac{1}{\beta} e^{\beta x} x^\nu K_\nu(x) + \frac{1}{\beta}\int_x^\infty e^{\beta t} t^\nu K_{\nu-1}(t)\,\mathrm{d}t. \tag{2.12}$$

Applying the inequality (A.8) and rearranging gives

$$\left(\frac{1}{\beta}-1\right)\int_x^\infty e^{\beta t} t^\nu K_\nu(t)\,\mathrm{d}t < \frac{1}{\beta} e^{\beta x} x^\nu K_\nu(x).$$

Inequality (2.10) for $\beta>0$ now follows on rearranging.

The case $\beta\le 0$ is simple. Since $e^{\beta t}$ is a non-increasing function of t when $\beta\le 0$ we have

$$\int_x^\infty e^{\beta t} t^\nu K_\nu(t)\,\mathrm{d}t \le e^{\beta x}\int_x^\infty t^\nu K_\nu(t)\,\mathrm{d}t < e^{\beta x} x^\nu K_\nu(x) \le \frac{1}{1-|\beta|} e^{\beta x} x^\nu K_\nu(x),$$

where we used inequality (2.9) to obtain the second inequality. Hence inequality (2.10)has been proved.

(iv) The case $\nu=1/2$ is simple. Using (A.1) we may easily integrate $t^{1/2}K_{1/2}(t)$:

$$\int_x^\infty t^{1/2}K_{1/2}(t)\,dt = \int_x^\infty \sqrt{\frac{\pi}{2}}e^{-t}\,dt = \sqrt{\frac{\pi}{2}}e^{-x} = x^{1/2}K_{1/2}(x).$$

It therefore follows that inequality (2.11) holds for $\nu=1/2$ because we have

$$\frac{\sqrt{\pi}\Gamma(1)}{\Gamma(1/2)} = 1,$$

where we used the facts that $\Gamma(1)=1$ and $\Gamma(1/2)=\sqrt{\pi}$.

Now suppose $\nu>1/2$. We begin by defining the function $u(x)$ to be

$$u(x) = Mx^\nu K_\nu(x) - \int_x^\infty t^\nu K_\nu(t)\,dt,$$

where

$$M = \frac{\sqrt{\pi}\Gamma(\nu+1/2)}{\Gamma(\nu)}.$$

We now show that $u(x)\geq 0$ for all $x\geq 0$, which will prove the result. We begin by noting that $\lim_{x\to 0+} u(x) = 0$ and $\lim_{x\to\infty} u(x) = 0$, which are verified by the following calculations, where we make use of the asymptotic formula (A.3) and the definite integral formula (A.17).

$$\begin{aligned} u(0) &= \lim_{x\to 0^+} \frac{\sqrt{\pi}\Gamma(\nu+1/2)}{\Gamma(\nu)}x^\nu K_\nu(x) - \int_0^\infty t^\nu K_\nu(t)\,dt \\ &= \sqrt{\pi}\Gamma(\nu+1/2)2^{\nu-1} - \sqrt{\pi}\Gamma(\nu+1/2)2^{\nu-1} \\ &= 0, \end{aligned}$$

and

$$\lim_{x\to\infty} u(x) = \lim_{x\to\infty} Mx^{\nu}K_{\nu}(x) - \lim_{x\to\infty}\int_{x}^{\infty} t^{\nu}K_{\nu}(t)\,\mathrm{d}t = 0,$$

where we used the asymptotic formula (A.5) to obtain the above equality. We may obtain an expression for the first derivative of u(x) by the use of the differentiation formula (A.13) as follows

$$u'(x) = x^{\nu}\left[K_{\nu}(x) - MK_{\nu-1}(x)\right]. \tag{2.13}$$

In the limit $x \to 0^{+}$ we have, by the asymptotic formula (A.3), that

$$u'(x) \sim \begin{cases} x^{\nu}\{2^{\nu-1}\Gamma(\nu)\frac{1}{x^{\nu}} - M2^{|\nu-1|-1}\Gamma(|\nu-1|)\frac{1}{x^{|\nu-1|}}\}, & \nu \neq 1, \\ x^{\nu}\{2^{\nu-1}\Gamma(\nu)\frac{1}{x^{\nu}} + M\log x\}, & \nu = 1. \end{cases}$$

Since $\nu > |\nu - 1|$ for $\nu > 1/2$ and $\lim_{x\to 0^+} x^{a}\log x = 0$, where $a > 0$, we have

$$u'(x) \sim 2^{\nu-1}\Gamma(\nu), \quad \text{as } x \to 0^{+}, \text{ for } \nu > 1/2.$$

Therefore u(x) is initially an increasing function of x. In the limit $x \to \infty$ we have, by(A.5),

$$u'(x) \sim \left(1 - \frac{\sqrt{\pi}\Gamma(\nu+1/2)}{\Gamma(\nu)}\right)\sqrt{\frac{\pi}{2}}x^{\nu-1/2}\mathrm{e}^{-x} < 0, \quad \text{for } \nu > 1/2.$$

We therefore see that u(x) is a decreasing function of x for large, positive x. From the formula (2.13) we see that x^{*} is a turning point of u(x) if and only if

$$K_{\nu}(x^{*}) = \frac{\sqrt{\pi}\Gamma(\nu+1/2)}{\Gamma(\nu)}K_{\nu-1}(x^{*}). \tag{2.14}$$

From Corollary 2.3, it follows that Eq. (2.14) has one root for $\nu>1/2$ (for which $\frac{\sqrt{\pi}\Gamma(\nu+1/2)}{\Gamma(\nu)}>1$).

Putting these results together, we see that $u(x)$ is non-negative at the origin and initially increases until it reaches it maximum value at x^*, it then decreases and tends to 0 as $x\to\infty$. Therefore $u(x)$ is non-negative for all $x\geq0$ when $\nu>1/2$.

(v) The proof for $\beta\leq0$ is easy and follows immediately from part (iv), since $1<\frac{2}{(1-\beta^2)^{\nu+1/2}}$ for $\nu\geq1/2$. So we suppose $\beta>0$. Again, because $K_{1/2}(x)=\sqrt{\frac{\pi}{2x}}\mathrm{e}^{-x}$, the case $\nu=1/2$ is straightforward, so we also suppose $\nu>1/2$. We make use of a similar argument to the one used in the proof of part (iv). We define the function $v(x)$ to be

$$v(x)=N\mathrm{e}^{\beta x}x^{\nu}K_{\nu}(x)-\int_x^{\infty}\mathrm{e}^{\beta t}t^{\nu}K_{\nu}(t)\,\mathrm{d}t,$$

where

$$N=\frac{2\sqrt{\pi}\Gamma(\nu+1/2)}{(1-\beta^2)^{\nu+1/2}\Gamma(\nu)}.$$

We now show that $v(x)\geq0$ for all $x\geq0$, which will prove the result. We begin by noting that $\lim_{x\to0^+}v(x)>0$ and $\lim_{x\to\infty}v(x)=0$, which are verified by the following calculations, where we make use of the asymptotic formula (A.3) and the definite integral formula (A.17).

$$v(0) = \lim_{x\to 0^+} \frac{2\sqrt{\pi}\Gamma(\nu+1/2)}{(1-\beta^2)^{\nu+1/2}\Gamma(\nu)} x^\nu K_\nu(x) - \int_0^\infty e^{\beta t} t^\nu K_\nu(t)\,dt$$

$$= \frac{2\sqrt{\pi}\Gamma(\nu+1/2)}{(1-\beta^2)^{\nu+1/2}\Gamma(\nu)} \cdot 2^{\nu-1}\Gamma(\nu) - \int_0^\infty e^{\beta t} t^\nu K_\nu(t)\,dt$$

$$> \frac{2\sqrt{\pi}\Gamma(\nu+1/2)}{(1-\beta^2)^{\nu+1/2}\Gamma(\nu)} \cdot 2^{\nu-1}\Gamma(\nu) - \int_{-\infty}^\infty e^{\beta t} |t|^\nu K_\nu(|t|)\,dt$$

$$= \frac{\sqrt{\pi}\Gamma(\nu+1/2)2^\nu}{(1-\beta^2)^{\nu+1/2}} - \frac{\sqrt{\pi}\Gamma(\nu+1/2)2^\nu}{(1-\beta^2)^{\nu+1/2}}$$

$$= 0,$$

and

$$\lim_{x\to\infty} v(x) = \lim_{x\to\infty} N e^{\beta x} x^\nu K_\nu(x) - \lim_{x\to\infty} \int_x^\infty e^{\beta t} t^\nu K_\nu(t)\,dt = 0,$$

where we used the asymptotic formula (A.5) to obtain the above equality. We may obtain an expression for the first derivative of v(x) by the use of the differentiation formula (A.13) as follows

$$v'(x) = e^{\beta x} x^\nu \left[(1+N\beta)K_\nu(x) - NK_{\nu-1}(x)\right]. \tag{2.15}$$

In the limit $x \to 0^+$ we have, by the asymptotic formula (A.3), that

$$v'(x) \sim \begin{cases} e^{\beta x} x^\nu \{2^{\nu-1}\Gamma(\nu)(1+N\beta)\frac{1}{x^\nu} - N \cdot 2^{|\nu-1|-1}\Gamma(|\nu-1|)\frac{1}{x^{|\nu-1|}}\}, & \nu \neq 1, \\ e^{\beta x} x^\nu \{2^{\nu-1}\Gamma(\nu)(1+N\beta)\frac{1}{x^\nu} + N\log x\}, & \nu = 1. \end{cases}$$

As in part (iv), we see that v(x) is initially an increasing function of *x* . In the limitx→∞ we have

$$v'(x) \sim (1 - N(1-\beta))\sqrt{\frac{\pi}{2}} x^{\nu-1/2} e^{(\beta-1)x}, \quad \text{for } \nu > 1/2.$$

Now, for $\nu > 1/2$ and $0 < \beta < 1$ we have, by (A.5),

$$N(1-\beta) = \frac{2\sqrt{\pi}\Gamma(\nu+1/2)}{\Gamma(\nu)} \cdot \frac{1}{(1-\beta^2)^{\nu-1/2}} \cdot \frac{1}{1+\beta} > 2 \cdot 1 \cdot \frac{1}{2} = 1. \tag{2.16}$$

Hence, $v(x)$ is a decreasing function of *x* for large, positive *x*. From formula (2.15) we see that x^* is a turning point of $v(x)$ if and only if

$$(1 + N\beta)K_\nu(x^*) = NK_{\nu-1}(x^*). \tag{2.17}$$

Inequality (2.16) shows that $N > 1 + N\beta$ for all $\nu > 1/2$ and $0 < \beta < 1$. From Corollary 2.3, it follows that Eq. (2.17) has one root for positive *x* and therefore $v(x)$ has one maximum which occurs at positive *x*. Putting these results together we see that $v(x)$ is positive at the origin and initially increases until it reaches it maximum value at x^*, it then decreases and tends to 0 as $x \to \infty$. Therefore $v(x)$ is non-negative for all $x \geq 0$ when $\nu > 1/2$, which completes the proof.

Combining the inequalities of Theorem 2.1 and Theorem 2.5 and the indefinite integral formula (1.1) we may obtain lower and upper bounds for the quantity $L_\nu(x)\mathbf{L}_{\nu-1}(x) - L_{\nu-1}(x)\mathbf{L}_\nu(x)$. Here is an example:

COROLLARY 2.6.

Suppose $\nu > -1/2$, *then for all* $x > 0$ *we have*

$$\frac{x^{\nu-1}I_{\nu+1}(x)}{\sqrt{\pi}2^{\nu-1}\Gamma(\nu+1/2)} < I_\nu(x)\mathbf{L}_{\nu-1}(x) - I_{\nu-1}(x)\mathbf{L}_\nu(x) < \frac{(\nu+1)x^{\nu-1}I_{\nu+1}(x)}{\sqrt{\pi}2^{\nu-1}\Gamma(\nu+3/2)}.$$

PROOF.

From the asymptotic formulas (A.2) and (A.6) for $I_\nu(x)$ and $\mathbf{L}(x)$, respectively, we have that

$$\lim_{x\downarrow 0}\left(x\left(I_\nu(x)\mathbf{L}_{\nu-1}(x)-I_{\nu-1}(x)\mathbf{L}_\nu(x)\right)\right)=0, \quad \text{for } \nu>-1/2.$$

Therefore, applying the indefinite integral formula (1.1) gives, for $\nu>-1/2$,

$$\int_0^x t^\nu I_\nu(t)\,\mathrm{d}t=\sqrt{\pi}2^{\nu-1}\Gamma(\nu+1/2)x\left(I_\nu(x)\mathbf{L}_{\nu-1}(x)-I_{\nu-1}(x)\mathbf{L}_\nu(x)\right). \tag{2.18}$$

From inequalities (2.2) and (2.6) of Theorem 2.1, we have

$$x^\nu I_{\nu+1}(x)<\int_0^x t^\nu I_\nu(t)\,\mathrm{d}t<\frac{2(\nu+1)}{2\nu+1}x^\nu I_{\nu+1}(x).$$

Substituting this inequality into (2.18) gives

$$\begin{aligned} x^\nu I_{\nu+1}(x) &< \sqrt{\pi}2^{\nu-1}\Gamma(\nu+1/2)x\left(I_\nu(x)\mathbf{L}_{\nu-1}(x)-I_{\nu-1}(x)\mathbf{L}_\nu(x)\right) \\ &< \frac{2(\nu+1)}{2\nu+1}x^\nu I_{\nu+1}(x). \end{aligned}$$

The desired inequality now follows from rearranging terms and an application of the standard formula $x\Gamma(x)=\Gamma(x+1)$.

REMARK 2.7.

The lower and upper bounds for $I_\nu(x)\mathbf{L}_{\nu-1}(x)-I_{\nu-1}(x)\mathbf{L}_\nu(x)$ that are given inCorollary 2.6 are simple, but very tight for large ν.

3. INEQUALITIES FOR THE MODIFIED BESSEL FUNCTION OF THE SECOND KIND

We now present some simple inequalities for the modified Bessel function of the second kind $K_\nu(x)$. The following theorem establishes an inequality for the modified Bessel function $K_\nu(x)$ that is useful in the study of Stein's method for Variance-Gamma distributions (see Gaunt [5]).

THEOREM 3.1.

Let $\nu>0$ *and* $x\geq 0$, *then*

$$\frac{1}{x^2}-\frac{x^{\nu-2}K_\nu(x)}{2^{\nu-1}\Gamma(\nu)} \tag{3.1}$$

is a monotone decreasing function of x on $(0,\infty)$ *and satisfies the following inequality*

$$0<\frac{1}{x^2}-\frac{x^{\nu-2}K_\nu(x)}{2^{\nu-1}\Gamma(\nu)}\leq\frac{1}{4(\nu-1)},\quad \textit{for } x\geq 0,\ \nu>1. \tag{3.2}$$

The lower bound is also valid for all $\nu>0$.

PROOF.

Applying the differentiation formula (A.14) gives

$$\frac{\mathrm{d}}{\mathrm{d}x}\left(\frac{1}{x^2}-\frac{x^{\nu-2}K_\nu(x)}{2^{\nu-1}\Gamma(\nu)}\right)$$

$$=-\frac{2}{x^3}-\frac{(\nu-2)x^{\nu-3}K_\nu(x)-\frac{1}{2}(K_{\nu-1}(x)+K_{\nu+1}(x))x^{\nu-2}}{2^{\nu-1}\Gamma(\nu)}. \tag{3.3}$$

Using (A.11) we may simplify the numerator as follows

$$
\begin{aligned}
&(\nu-2)K_\nu(x)-\frac{1}{2}x\left(K_{\nu-1}(x)+K_{\nu+1}(x)\right)\\
&\quad=(\nu-2)K_\nu(x)-\frac{1}{2}x\left(2K_{\nu-1}(x)+\frac{2\nu}{x}K_\nu(x)\right)\\
&\quad=-xK_{\nu-1}(x)-2K_\nu(x).
\end{aligned}
$$

Hence, (3.3) simplifies to

$$\frac{\mathrm{d}}{\mathrm{d}x}\left(\frac{1}{x^2}-\frac{x^{\nu-2}K_\nu(x)}{2^{\nu-1}\Gamma(\nu)}\right)=\frac{-2^\nu\Gamma(\nu)+x^{\nu+1}K_{\nu-1}(x)+2x^\nu K_\nu(x)}{2^{\nu-1}\Gamma(\nu)x^3}.$$

Thus, proving that (3.1) is monotone decreasing reduces to proving that, for $x>0$,

$$x^{\nu+1}K_{\nu-1}(x)+2x^\nu K_\nu(x)<2^\nu\Gamma(\nu). \tag{3.4}$$

From (A.13) we get that

$$
\begin{aligned}
\frac{\mathrm{d}}{\mathrm{d}x}\left(x^{\nu+1}K_{\nu-1}(x)+2x^\nu K_\nu(x)\right)&=\frac{\mathrm{d}}{\mathrm{d}x}\left(x^2\cdot x^{\nu-1}K_{\nu-1}(x)+2x^\nu K_\nu(x)\right)\\
&=2x^\nu K_{\nu-1}(x)-x^{\nu+1}K_{\nu-2}(x)-2x^\nu K_{\nu-1}(x)\\
&=-x^{\nu+1}K_{\nu-2}(x)\\
&<0.
\end{aligned}
$$

So $x^{\nu+1}K_{\nu-1}(x)+2x^\nu K_\nu(x)$ is a monotone decreasing function of x and from the asymptotic formula (A.3) we see that its limit as $x\to0^+$ is $\lim_{x\to0^+}(x^{\nu+1}K_{\nu-1}(x)+2x^\nu K_\nu(x))=2\cdot2^{\nu-1}\Gamma(\nu)=2^\nu\Gamma(\nu)$. Therefore (3.4) is proved, and so (3.1) is monotone decreasing on $(0,\infty)$. It is therefore bounded above and below its values in the limits $x\to\infty$ and $x\to0$. These are calculated using the asymptotic formulas (A.5)and (A.4) and are given below:

$$\lim_{x\to\infty}\left(\frac{1}{x^2}-\frac{x^{\nu-2}K_\nu(x)}{2^{\nu-1}\Gamma(\nu)}\right)=0,$$

$$\lim_{x\to 0+}\left(\frac{1}{x^2}-\frac{x^{\nu-2}K_\nu(x)}{2^{\nu-1}\Gamma(\nu)}\right)=\frac{2^{\nu-3}\Gamma(\nu-1)}{2^{\nu-1}\Gamma(\nu)}=\frac{1}{4(\nu-1)},$$

where the first limit holds for all $\nu>0$ and the second limit is valid for all $\nu>1$. This completes the proof.

REMARK 3.2.

Inequality (3.2) of Theorem 3.1 is closely related to some inequalities given by Ismail [8]and Baricz et al. [3]. Ismail proved that $x^\nu K_\nu(x)e^x>2^{\nu-1}\Gamma(\nu)$ for $x>0$, $\nu>1/2$, and Baricz et al. proved that $x^{\nu-1}K_\nu(x)\geq 2^{\nu-1}\Gamma(\nu)K_1(x)$ for $x>0$, $\nu\geq 1$, which improves on the bound of Ismail for all $\nu\geq 1$. From inequality (3.2) we can obtain lower and upper bounds for the quantity $x^\nu K_\nu(x)$. The upper bound is $x^\nu K_\nu(x)<2^{\nu-1}\Gamma(\nu)$ for $x>0$,$\nu>0$, and therefore, for $x>0$,

$$2^{\nu-1}\Gamma(\nu)e^{-x}<x^\nu K_\nu(x)<2^{\nu-1}\Gamma(\nu),\quad \nu>0,$$

which can be improved as follows when $\nu\geq 1$:

$$2^{\nu-1}\Gamma(\nu)e^{-x}<2^{\nu-1}\Gamma(\nu)xK_1(x)\leq x^\nu K_\nu(x)<2^{\nu-1}\Gamma(\nu).$$

Finally, we establish a simple, but surprisingly tight, lower bound for the modified Bessel function $K_0(x)$.

THEOREM 3.3.

Let $x>0$, *then*

$$\frac{\Gamma(x+1/2)}{\Gamma(x+1)}<\sqrt{\frac{2}{\pi}}e^x K_0(x). \tag{3.5}$$

PROOF.

Formula 10.32.8 of Olver et al. [13] gives the following integral representation of $K_0(x)$:

$$K_0(x) = \int_1^\infty \frac{e^{-xt}}{\sqrt{t^2-1}}\,dt, \quad x > 0.$$

Setting $t=2u+1$ gives

$$K_0(x) = e^{-x} \int_0^\infty \frac{e^{-2xu}}{\sqrt{u^2+u}}\,du.$$

For $u>0$ we have $e^{2u} - 1 = \sum_{k=1}^\infty \frac{(2u)^k}{k!} > 2u + 2u^2$, and so

$$e^x K_0(x) > \sqrt{2} \int_0^\infty \frac{e^{-2xu}}{\sqrt{e^{2u}-1}}\,du = \sqrt{2} \int_0^\infty \frac{e^{-(2x+1)u}}{\sqrt{1-e^{-2u}}}\,du, \quad \text{for } x > 0.$$

Making the change of variables $y=e^{-2u}$ gives

$$\begin{aligned}
\sqrt{2} \int_0^\infty \frac{e^{-(2x+1)u}}{\sqrt{1-e^{-2u}}}\,du &= \frac{1}{\sqrt{2}} \int_0^1 (1-y)^{-1/2} y^{x-1/2}\,dy \\
&= \frac{1}{\sqrt{2}} B(1/2, x+1/2) \\
&= \frac{\Gamma(1/2)\Gamma(x+1/2)}{\sqrt{2}\Gamma(x+1)} \\
&= \frac{\sqrt{\pi}\Gamma(x+1/2)}{\sqrt{2}\Gamma(x+1)},
\end{aligned}$$

where B(a,b) is the beta function, and we used the standard formula $B(a,b)=\frac{\Gamma(a)\Gamma(b)}{\Gamma(a+b)}$ to obtain the third equality. This completes the proof.

COROLLARY 3.4.

Let x>0, *then*

$$\frac{1}{\sqrt{x+1/2}}<\sqrt{\frac{2}{\pi}}e^{x}K_0(x)<\frac{1}{\sqrt{x}}.$$

PROOF.

The upper bound follows because $K_0(x)<K_{1/2}(x)=\sqrt{\frac{\pi}{2x}}e^{-x}$.

The lower bound follows since $\frac{\Gamma(x+1/2)}{\Gamma(x+1)}>\frac{1}{\sqrt{x+1/2}}$, which we now prove. Examining the proof of Theorem 3.3 we see that

$$\frac{\Gamma(x+1/2)}{\Gamma(x+1)}=\frac{2}{\sqrt{\pi}}\int_0^\infty \frac{e^{-(2x+1)u}}{\sqrt{1-e^{-2u}}}du.$$

Now, for u>0 we have $1-e^{-2u}=\sum_{k=1}^{\infty}(-1)^{k+1}\frac{(2u)^k}{k!}<2u$, and so

$$\frac{\Gamma(x+1/2)}{\Gamma(x+1)}>\frac{2}{\sqrt{\pi}}\int_0^\infty \frac{e^{-(2x+1)u}}{\sqrt{2u}}du=\frac{2\sqrt{2}}{\sqrt{\pi}}\int_0^\infty e^{-(2x+1)v^2}dv=\frac{1}{\sqrt{x+1/2}},$$

as required.

REMARK 3.5.

Luke [11] obtained the following bounds for $K_0(x)$:

$$\frac{8\sqrt{x}}{8x+1} < \sqrt{\frac{2}{\pi}} e^{x} K_0(x) < \frac{16x+7}{(16x+9)\sqrt{x}}.$$

Numerical experiments show that the bounds of Luke and our lower bound of Corollary 3.3 are remarkably accurate for all but very small x, for which the logarithmic singularity of $K_0(x)$ blows up. The lower bound $\frac{8\sqrt{x}}{8x+1}$ outperforms our bound lower bound of $\frac{\Gamma(x+1/2)}{\Gamma(x+1)}$ for $x>0.394$ (3 d.p.), whilst our bound outperforms for $x<0.394$ (3 d.p.), and performs considerably better for very small x.

ACKNOWLEDGMENTS

During the course of this research the author was supported by an EPSRC DPhil Studentship, and an EPSRC Doctoral Prize. The author would like to thank Gesine Reinert for some fruitful discussions. Finally, the author would like to thank an anonymous referee for his/her helpful comments and suggestions.

REFERENCES

1. Á. Baricz, Turán type inequalities for modified Bessel functions, Bull. Aust. Math. Soc. 82 (2010) 254–264.

2. Á. Baricz, Bounds for modified Bessel functions of the first and second kinds, Proc. Edinb. Math. Soc. 53 (2010) 575–599.

3 . Á. Baricz, D. Jankov, T.K. Pogány, Turán type inequalities for Krätzel functions, J. Math. Anal. Appl. 388 (2012) 716–724.

4. Á. Baricz, Y. Sun, Bounds for the generalized Marcum Q-function, Appl. Math. Comput. 217 (2010) 2238–2250.

5. R.E. Gaunt, Rates of convergence of Variance-Gamma approximations via Stein's method, DPhil thesis, University of Oxford, 2013.

6. I.S. Gradshetyn, I.M. Ryzhik, Table of Integrals, Series and Products, seventh ed., Academic Press, 2007.

7. E.K. Ifantis, P.D. Siafarikas, Inequalities involving Bessel and modified Bessel functions, J. Math. Anal. Appl. 147 (1990) 214–227.

8. M.E.H. Ismail, Complete monotonicity of modified Bessel functions, Proc. Amer. Math. Soc. 108 (1990) 353–361.

9. M.E.H. Ismail, M.E. Muldoon, Monotonicity of the zeros of a cross-product of Bessel functions, SIAM J. Math. Anal. 9 (1978) 759–767.

10. A.L. Jones, An extension of an inequality involving modified Bessel functions, J. Math. Phys. 47 (1968) 220–221.

11. Y.L. Luke, Inequalities for generalized hypergeometric functions, J. Approx. Theory 5 (1972) 41–65.

12. Nåsell, Inequalities for modified Bessel functions, Math. Comp. 28 (1974) 253–256.

13. F.W.J. Olver, D.W. Lozier, R.F. Boisvert, C.W. Clark, NIST Handbook of Mathematical Functions, Cambridge University Press, 2010.

14. J. Segura, Bounds for ratios of modified Bessel functions and associated Turán-type inequalities, J. Math. Anal. Appl. 374 (2011) 516–528.

CHAPTER 7

A Note on the Almost Sure Central Limit Theorem for Partial Sums of ρ⁻-Mixing Sequences

Feng Xu, Qunying Wu

College of Science, Guilin University of Technology, Guilin, China

ABSTRACT

Let $\{X_n\}_{n\in N}$ be a strictly stationary sequence of ρ^--mixing random variables. We proved the almost sure central limit theorem, containing the general weight sequences, for the partial sums S_n/σ_n, where $S_n=\sum_{i=1}^{n}X_i$, $\sigma_n^2=ES_n^2$. The result generalizes and improves the previous results.

KEYWORDS

ρ⁻-Mixing Sequences, Partial Sums, Almost Sure Central Limit Theorem

1. INTRODUCTION

Let $\mathscr{C}$ be a class of functions which are coordinatewise increasing. For a random variable X, define

$$\|X\|_p = \left(\mathrm{E}|X|^p\right)^{1/p}.$$

For two nonempty disjoint sets $S,T \subset N$, we define $\text{dist}(S,T)$ to be $\min\{|j-k|; j \in S, k \in T\}$. Let $\sigma(S)$ be the σ-field generated by $\{X_k, k \in S\}$, and define $\sigma(T)$ similarly.

A sequence $\{X_n, n \geq 1\}$ is called negatively associated (NA) if for ever pair of disjoint subsets S, T of N,

$$\text{cov}\{f(X_i, i \in S), g(X_j, j \in T)\} \leq 0,$$

where $f, g \in \mathscr{C}$. $\{X_n, n \geq 1\}$ is called ρ*-mixing, if

$$\rho^*(k) = \sup\{\rho(S,T); S,T \subset N, \text{dist}(S,T) \geq k\} \to 0, k \to \infty,$$

Where

$$\rho(S,T)=\sup\left\{\frac{|E(f-Ef)(g-Eg)|}{\|f-Ef\|_2\cdot\|g-Eg\|_2};f\in L_2(\sigma(S)),f\in L_2(\sigma(T))\right\}.$$

Definition 1. [1] A sequence $\{X_n, n\geq 1\}$ is called ρ⁻-mixing, if

$$\rho^-(k)=\sup\{\rho^-(S,T):\text{dist}(S,T)\geq k, S,T\subset N\}\to 0,\text{ as } k\to\infty,$$

Where

$$\rho^-(S,T)=0\vee\sup\left\{\frac{\text{cov}\{f(X_i,i\in S),g(X_j,t\in T)\}}{\sqrt{\text{var}\{f(X_i,i\in S)\}\text{var}\{g(X_j,t\in T)\}}};f,g\in\mathscr{C}\right\}.$$

The definition of NA is given by Joag-Dev and Proschan [2] , and the concept of ρ*-mixing random variables is given by Kolmogorov and Rozanov [3] . In 1999, the concept of ρ⁻-mixing random variables was introduced initially by Zhang and Wang [1] . Obviously, ρ⁻-mixing random variables include NA and ρ*-mixing random variables, which have a lot of applications. Their limit properties have received more and more attention recently, and a number of results have been obtained, such as Zhang and Wang [1] for Rosenthal-type moment inequality and Marcinkiewicz-Zygmund law of large numbers, Zhang [4] for the central limit theorems of random fields, Wang and Lu [5] for the weak convergence theorems.

Starting with Brosamler [6] and Schatte [7] , in the last two decades several authors investigated the almost sure central limit theorem (ASCLT) for partial sums S_n/σ_n of random variables. We refer the reader to Brosamler [6] , Schatte [7] , Lacey and Philipp [8] , Ibragimov and Lifshits [9] , Berkes and Csáki [10] , Hörmann [11] and Wu [12] . The simplest form of the ASCLT [6] - [8] reads as follows: let $\{X_n; n \geq 1\}$ be i.i.d. random variables with mean 0, variance $\sigma^2 > 0$ and partial sums S_n. Then

$$\lim_{n\to\infty} \frac{1}{\log n} \sum_{k=1}^{n} \frac{1}{k} I\left(\frac{S_k}{\sigma\sqrt{k}} \leq x\right) = \Phi(x) \text{ a.s. for any } x \in R. \tag{1}$$

where I denotes indicator function, and $\Phi(x)$ is the standard normal distribution function. For other version of ρ⁻-mixing sequences, see [13] -[15] .

The purpose of this article is to study and establish the ASCLT, containing the general weight sequences, for partial sums of ρ⁻-mixing sequence. Our results not only generalize and improve those on ASCLT previously obtained by Brosamler [6] , Schatte [7] and Lacey and Philipp [8] from the i.i.d. case to ρ⁻-mixing sequences, but also expand the scope of the weights from $1/k$ to $\exp(\log^{\alpha} k)/k$, $0 \leq \alpha < 1/2$.

Throughout this paper, $a_n \sim b_n$ means $\lim_{n\to\infty} a_n/b_n = 1$; and set the positive absolute constant c to vary from line to line.

Theorem 1. Let $\{X_n\}_{n\in N}$ be a strictly stationary ρ⁻-mixing sequence with $EX_1 = 0$, $0 < E|X_1|^r < \infty$ for a certain $r > 2$, and denote $S_n = \sum_{i=1}^{n} X_i$, $\sigma_n^2 = ES_n^2$. Assume that

(a) $\sigma^2 = EX_1^2 + 2\sum_{k=2}^{\infty} \text{cov}(X_1, X_k) > 0,$

(b) $\sum_{k=2}^{\infty}\left|\operatorname{cov}(X_1, X_k)\right| < \infty,$

(c) $\sum_{k=1}^{\infty}\frac{\rho^-(k)}{k} < \infty.$

Suppose $0 \le \alpha < 1/2$ and set

$$d_k = \frac{\exp\left(\log^{\alpha} k\right)}{k}, D_n = \sum_{k=1}^{n} d_k. \tag{2}$$

then

$$\lim_{n\to\infty}\frac{1}{D_n}\sum_{k=1}^{n} d_k I\left(\frac{S_k}{\sigma_n} \le x\right) = \Phi(x) \text{ a.s. for any } x \in \mathrm{R}. \tag{3}$$

Remark 1. By the terminology of summation procedures (cf. [16] , p. 35), Theorem 1 remains valid if we replace the weight sequence $\{d_k\}_{k\ge1}$ by any $\{d_k^*\}_{k\ge1}$ such that $0 \le d_k^* \le d_k$ and $\sum_{k\ge1} d_k^* = \infty$.

Remark 2. ρ⁻-mixing random variables include NA and ρ*-mixing random variables, so for NA and ρ*-mixing random variables sequences Theorem 1 also holds.

Remark 3. Essentially, the open problem that whether Theorem 1 holds for $1/2 \le \alpha < 1$ still remains open.

2. SOME LEMMAS

Lemma 1. [4] Let $\{X_n, n \geq 1\}$ be a weakly stationary ρ⁻-mixing sequence with $EX_n = 0$, $0 < EX_1^2 < \infty$, and

$$\sigma^2 = EX_1^2 + 2\sum_{k=2}^{\infty} \text{cov}(X_1, X_k) > 0, \sum_{k=2}^{\infty} |\text{cov}(X_1, X_k)| < \infty,$$

then

$$\frac{\sigma_n^2}{n} \to \sigma^2, \quad \frac{S_n}{\sigma_n} \xrightarrow{d} \mathcal{N}, \text{ as } n \to \infty,$$

where $\mathcal{N}$ denotes the standard normal random variable.

Lemma 2. [5] For a positive real number $q \geq 2$, if $\{X_n, n \geq 1\}$ is a sequence of ρ⁻-mixing random variables with $EX_i = 0$, $E|X_i|^q < \infty$ for every $i \geq 1$, then for all $n \geq 1$, there is a positive constant $c = c\left(q, \rho^-(\cdot)\right)$ such that

$$E\left(\max_{1 \leq j \leq n} |S_j|^q\right) \leq c\left(\sum_{i=1}^{n} E|X_i|^q + \left(\sum_{i=1}^{n} EX_i^2\right)^{q/2}\right).$$

Lemma 3. [17] Let $\{X_n, n \geq 1\}$ be a weakly stationary ρ⁻-mixing sequence. Assume $\sup_n E|X_n|^r < \infty.$ Then for any bounded Lipschitz function f: $R \to R$, We have

$$\left|\operatorname{cov}\left(f\left(\frac{S_i}{\sigma_i}\right), f\left(\frac{S_j}{\sigma_j}\right)\right)\right| \le c\left|-\frac{1}{\sigma_i\sigma_j}\sum_{l=1}^{i}\sum_{m=2i+1}^{2i+l}\operatorname{cov}(X_l, X_m) + 8\rho^-(i) + 2\frac{\sigma_{2i}}{\sigma_j}\right|$$

Lemma 4. Let $\{\xi, \xi_n\}_{n\in N}$ be a sequence of uniformly bounded random variables. Assume that

$\sum_{k=1}^{\infty}\frac{\rho^-(k)}{k} < \infty,$ and existing constants $c > 0$ and $\varepsilon > 0$ such that

$$|\mathrm{E}\xi_k\xi_l| \le c\left(\rho^-(k) + \left(\frac{k}{l}\right)^{\varepsilon}\right), \text{ for } 1 \le 2k < l,$$

then

$$\lim_{n\to\infty}\frac{1}{D_n}\sum_{k=1}^{n} d_k\xi_k = 0 \text{ a.s.,} \tag{4}$$

where d_k and D_n are defined by (2).

Proof. Set $T_n = \frac{1}{D_n}\sum_{k=1}^{n} d_k\xi_k$, we get

$$\begin{aligned}\mathrm{E}T_n^2 &= \frac{1}{D_n^2}\mathrm{E}\left(\sum_{k=1}^{n} d_k\xi_k\right)^2 \le \frac{1}{D_n^2}\sum_{1\le k\le l\le n, 2k\ge l} d_k d_l|\mathrm{E}\xi_k\xi_l| + \frac{1}{D_n^2}\sum_{1\le k\le l\le n, 2k<l} d_k d_l|\mathrm{E}\xi_k\xi_l| \\ &:= \frac{1}{D_n^2}T_{n1} + \frac{1}{D_n^2}T_{n2}.\end{aligned}$$

Firstly we estimate T_{n1}. Since ξ_k is a bounded random variable, we get

$$T_{n1} \le c\sum_{k=1}^{n}\sum_{l=k}^{2k} d_k d_l \le c\exp\left(\log^{\alpha} n\right)\sum_{k=1}^{n} d_k \sum_{l=k}^{2k}\frac{1}{l} \le cD_n \exp\left(\log^{2} n\right).$$

Now we estimate T_{n2}. By the conditions $|\mathrm{E}\xi_k\xi_l| \le c\left(\rho^{-}(k)+\left(\frac{k}{l}\right)^{\varepsilon}\right)$ for $l > 2k$, we get

$$T_{n2} = \sum_{1\le k\le l\le n, 2k<l} d_k d_l |\mathrm{E}\xi_k\xi_l| \le c\sum_{l=2}^{n}\sum_{k=1}^{l-1} d_k d_l \rho^{-}(k) + c\sum_{l=2}^{n}\sum_{k=1}^{l-1} d_k d_l \left(\frac{k}{l}\right)^{\varepsilon}$$

$$\le c\sum_{l=1}^{n} d_l \sum_{k=1}^{n} d_k \rho^{-}(k) + c\sum_{l=1}^{n}\sum_{k=1}^{l} d_k d_l \left(\frac{k}{l}\right)^{\varepsilon} := A_1 + A_2.$$

By condition $\sum_{k=1}^{\infty}\frac{\rho^{-}(k)}{k} < \infty$, we obtain

$$A_1 \le c\exp\left(\log^{\alpha} n\right)\sum_{l=1}^{n} d_l \sum_{k=1}^{n}\frac{\rho^{-}(k)}{k} \le cD_n \exp\left(\log^{\alpha} n\right).$$

and

$$A_2 \le c\sum_{l=2}^{n}\sum_{k=1}^{l}\frac{\exp\left(\log^{\alpha} l\right)}{l^{1+\varepsilon}}\cdot\frac{\exp\left(\log^{\alpha} k\right)}{k^{1-\varepsilon}} \le c\exp\left(\log^{\alpha} n\right)\sum_{l=1}^{n}\frac{\exp\left(\log^{\alpha} l\right)}{l^{1+\varepsilon}}\cdot\frac{l^{\varepsilon}}{\varepsilon} \le cD_n \exp\left(\log^{\alpha} n\right).$$

Since $D_n \sim \frac{1}{\alpha}\log^{1-\alpha} n \exp\left(\log^{\alpha} n\right)$ and $\log D_n \sim \log^{\alpha} n$ for $0<\alpha<1/2$ from the proof of Lemma 2.2 in

Wu [18] , we have, as $n \to \infty$,

$$\exp\left(\log^{\alpha} n\right) \sim \frac{\alpha D_n}{\left(\log D_n\right)^{(1-\alpha)/\alpha}} \sim \frac{\alpha D_n}{\log^{1-\alpha} n},$$

Thus

$$\mathrm{E}T_n^2 \le \frac{1}{D_n^2}\left(T_{n1} + A_1 + A_2\right) = c\frac{\exp\left(\log^\alpha n\right)}{D_n} \le \frac{c}{\log^{1-\alpha} n}.$$

Let $n_k = \exp\left(k^\tau\right)$, $\tau > 1/(1-\alpha)$, we get

$$\sum_{K=1}^{\infty} P\left(\left|T_{n_k}\right| > \varepsilon\right) \le c\sum_{k=1}^{\infty}\mathrm{E}T_{n_k}^2 \le c\sum_{k=1}^{\infty}\frac{1}{k^{(1-\alpha)\tau}} < \infty.$$

By Borel-Cantelli lemma,

$$T_{n_k} \to 0 \text{ a.s.}, k \to \infty.$$

For any n, existing n_k and n_{k+1} such that $n_k < n \le n_{k+1}$, then, by $\left|\xi_i\right| \le c$ for any i,

$$\left|T_n\right| \le \left|\frac{1}{D_{n_k}}\sum_{i=1}^{n_k} d_i\xi_i\right| + \frac{1}{D_{n_k}}\sum_{i=n_k+1}^{n_{k+1}} d_i\left|\xi_i\right| \le \left|T_{n_k}\right| + c\left(\frac{D_{n_{k+1}} - D_{n_k}}{D_{n_k}}\right) \to 0 \text{ a.s. } n \to \infty,$$

from $\frac{D_{n_{k+1}}}{D_{n_k}} \sim \frac{\exp\left((k+1)^\tau\right)}{\exp\left(k^\tau\right)} \sim \exp\left((k+1)^\tau\left(1-\left(\frac{k}{k+1}\right)^\tau\right)\right) \to 1$. i.e., (4) holds. This completes the proof of Lemma 4.

3. PROOF

Proof of Theorem 1. By Lemma 1, we have

$$\frac{S_k}{\sigma_k} \xrightarrow{d} \mathcal{N}, \text{ as } k \to \infty.$$

This implies that for any $g(x)$ which is a bounded function with bounded continuous derivatives,

$$\mathrm{E}g\left(\frac{S_k}{\sigma_k}\right) \to \mathrm{E}g(\mathcal{N}), \text{ as } k \to \infty,$$

Hence, by the Toeplitz lemma, we obtain

$$\lim_{n\to\infty} \frac{1}{D_n} \sum_{k=1}^{n} d_k \mathrm{E}g\left(\frac{S_k}{\sigma_k}\right) = \mathrm{E}g(\mathcal{N}).$$

In the other hand, from Theorem 7.1 of Billingsley [19] and Section 2 of Peligrad and Shao [20] , we know that (3) is equivalent to

$$\lim_{n\to\infty} \frac{1}{D_n} \sum_{k=1}^{n} d_k g\left(\frac{S_k}{\sigma_k}\right) = \mathrm{E}g(\mathcal{N}) \text{ a.s..}$$

Hence, to prove (3), it suffices to prove

$$\lim_{n\to\infty} \frac{1}{D_n} \sum_{k=1}^{n} d_k \left(g\left(\frac{S_k}{\sigma_k}\right) - \mathrm{E}g\left(\frac{S_k}{\sigma_k}\right)\right) = 0 \text{ a.s.,} \tag{5}$$

for any $g(x)$ which is a bounded function with bounded continuous derivatives.

Let $k \geq 1$, define

$$\xi_k = g\left(\frac{S_k}{\sigma_k}\right) - \mathrm{E}g\left(\frac{S_k}{\sigma_k}\right).$$

For any $1 \le 2k < l$, we get,

$$\begin{aligned}|\mathrm{E}\xi_k\xi_l| &= \left|\mathrm{cov}\left(g\left(\frac{S_k}{\sigma_k}\right), g\left(\frac{S_l}{\sigma_l}\right)\right)\right| \\ &= \left|\mathrm{cov}\left(g\left(\frac{S_k}{\sigma_k}\right), g\left(\frac{S_l}{\sigma_l}\right) - g\left(\frac{\sum_{i=2k+1}^{l} X_i}{\sigma_l}\right)\right)\right| + \left|\mathrm{cov}\left(g\left(\frac{S_k}{\sigma_k}\right), g\left(\frac{\sum_{i=2k+1}^{l} X_i}{\sigma_l}\right)\right)\right| \\ &:= I_1 + I_2.\end{aligned} \tag{6}$$

Firstly we estimate I_1. By Lemma 1 $\frac{\sigma_n^2}{n} \to \sigma^2$, we note that certain $n_0 \in N$, $0 < \varepsilon < \sigma$ exist such that $\frac{1}{\sigma_n} \le \frac{1}{(\sigma - \varepsilon)\sqrt{n}}$ as $n > n_0$. Since g is a bounded Lipschitz function, i.e., there exists a constant c > 0 such that

$|g(x)| \le c$, $|g(x) - g(y)| \le c|x - y|$ for any $x, y \in \mathrm{R}$. By Jensen inequality, Lemma 2 and $\sigma < \infty$, we obtain that

$$I_1 \le c\frac{\mathrm{E}\left|\sum_{i=1}^{2k} X_i\right|}{\sqrt{l}} \le c\frac{\sqrt{\mathrm{E}\left(\sum_{i=1}^{2k} X_i\right)^2}}{\sqrt{l}} \le c\frac{\sqrt{\left(\sum_{i=1}^{2k} \mathrm{E}X_i^2\right)}}{\sqrt{l}} \le c\frac{\sqrt{\left(\sum_{i=1}^{2k} \mathrm{E}X_1^2\right)}}{\sqrt{l}} \le c\left(\frac{k}{l}\right)^{1/2}. \tag{7}$$

Now we estimate I_2. Note that g is a bounded function with bounded continuous derivatives, so, by Lemma 3, we have

$$I_2 \le c\rho^-(k). \tag{8}$$

So if $l > 2k$, combining with (6), (7), (8), we obtain

$$\left|\mathrm{E}\xi_k\xi_l\right| \le c\left(\left(\frac{k}{l}\right)^{1/2} + \rho^-(k)\right).$$

By Lemma 4, (5) holds.

This completes the proof of Theorem 1.1.

ACKNOWLEDGMENTS

We thank the editor and the referee for their comments. This work is supported by National Natural Science Foundation of China (11361019).

REFERENCES

1. Zhang, L.X. and Wang, X.Y. (1999) Convergence Rates in the Strong Laws of Asymptotically Negatively Associated Random Fields. Applied Mathematics—A Journal of Chinese Universities Series B, 14, 406-416.

2. Joag-Dev, K. and Proschan, F. (1983) Negative Association of Random Variables with Applications. Annals of Statistics, 11, 286-295.

3. Kolmogorov, A.N. and Rozanov, U.A. (1960) On Strong Mixing Conditions for Stationary Gaussian Processes. Theory of Probability and Its Applications, 5, 204-208.

4. Zhang, L.X. (2000) Central Limit Theorems for Asymptotically Negatively Associated Random Fields. Acta Mathematica Sinica, 6, 691-710.

5. Wang, J.F. and Lu, F.B. (2006) Inequalities of Maximum of Partial Sums and Weak Convergence for a Class of Weak Dependent Random Variables. Acta Mathematica Sinica, 22, 693-700.

6. Brosamler, G.A. (1988) An Almost Everywhere Central Limit Theorem. Mathematical Proceedings of the Cambridge Philosophical Society, 104, 561-574.

7. Schatte, P. (1988) On Strong Versions of the Central Limit Theorem. Mathematische Nachrichten, 137, 249-256.

8. Lacey, M.T. and Philipp, W. (1990) A Note on the Almost Sure Central Limit Theorem. Statistics and Probability Letters, 9, 201-205.

9. Ibragimov, I.A. and Lifshits, M. (1998) On the Convergence of Generalized Moments in Almost Sure Central Limit Theorem. Statistics and Probability Letters, 40, 343-351.

10. Berkes, I. and Csáki, E. (2001) A Universal Result in Almost Sure Central Limit Theory. Stochastic Processes and Their Applications, 94, 105-134.

11. Hörmann, S. (2007) Critical Behavior in Almost Sure Central Limit Theory. Journal of Theoretical Probability, 20, 613-636.

12. Wu, Q.Y. (2011) Almost Sure Limit Theorems for Stable Distribution. Statistics and Probability Letters, 281, 662-672.

13. Zhang, M.D., Tan, X.L. and Zhang, Y. (2015) An Extension of Almost Sure Central Limit Theorem for Product of Partial Sums of ρ−-Mixing Sequences. Journal of Beihua University (Natural Science), 16, 427-430.

14. Tan, L.X., Zhang, Y. and Zhang, Y. (2012) An Almost Sure Central Limit Theorem of Products of Partial Sums for ρ--Mixing Sequences. Journal of Inequalities and Applications, 2012, 51-63.

15. Zhou, G.Y. and Zhang, Y. (2014) Almost Sure Central Limit Theorem of Products of Sums of Partial Sums for ρ--Mixing Sequences. Journal of Jilin University (Science Edition), 50, 1129-1134.

16. Chandrasekharan, K. and Minakshisundaram, S. (1952) Typical Means. Oxford University Press, Oxford.

17. Zhou, H. (2005) A Note on the Almost Sure Central Limit Theorem for ρ--Mixing Sequences. Chinese Journal Zhejiang University (Science Edition), 32, C503-C505.

18. Wu, Q.Y. (2012) A Note on the Almost Sure Limit Theorem for Self-Normalized Partial Sums of Random Variables in the Domain of Attraction of Thenormal Law. Journal of Inequalities and Applications, 2012, 17-26.

19. Billingsley, P. (1968) Convergence of Probability Measures. Wiley, New York.

20. Peligrad, M. and Shao, Q.M. (1995) A Note on the Almost Sure Central Limit Theorem for Weakly Dependent Random Variables. Statistics and Probability Letters, 22, 131-136.

CHAPTER 8

Frustration index and Cheeger inequalities for discrete and continuous magnetic Laplacians

Carsten Lange[1,2] · Shiping Liu[3] · Norbert Peyerimhoff[3] · Olaf Post[4]

1 Fachbereich für Mathematik und Informatik, Freie Universität Berlin, 14195 Berlin, Germany
2 Zentrum Mathematik, Technische Universität München, 85748 Garching, Germany
3 Department of Mathematical Sciences, Durham University, Durham DH1 3LE, UK
4 Fachbereich IV-Mathematik, Universität Trier, 54286 Trier, Germany

ABSTRACT

We discuss a Cheeger constant as a mixture of the frustration index and the expansion rate, and prove the related Cheeger inequalities and higher order Cheeger inequalities for graph Laplacians with cyclic signatures, discrete magnetic Laplacians on finite graphs and magnetic Laplacians on closed Riemannian manifolds. In this process, we develop spectral clustering algorithms for partially oriented graphs and multi-way spectral clustering algorithms via metrics in lens spaces and complex projective spaces. As a byproduct, we give a unified viewpoint of Harary's structural balance theory of signed graphs and the gauge invariance of magnetic potentials.

1 INTRODUCTION

Cheeger's inequality is one of the most fundamental and important estimates in spectral geometry. It was first proved by Cheeger for the Laplace-Beltrami operator on a Riemannian manifold [7] and later extended to the setting of discrete graphs, see e.g., [1, 2, 6, 11], demonstrating the close relationship between the spectrum and the geometry of the underlying space. This inequality has a tremendous impact in discrete and continuous theories and is an important intersection point for interactions between both communities. For example, it stimulated research in discrete mathematics such as spectral clustering algorithms for data mining [36], or the construction of expander graphs [25]. Cheeger inequalities have also been considered on metric graphs, see, e.g., [42] and, using a coarea formula in the proof [44]. We recently witness several fruitful interactions in the other direction: Lee, Oveis Gharan and Trevisan's higher order Cheeger inequalities [28, 29] on finite graphs were used by Miclo [40] to prove that hyperbounded, ergodic, and self-adjoint Markov operators admit a spectral gap, solving a 40-year-old conjecture of Simon and Høegh–Krohn [50]. For further developments, see [32, 57]. Another example is an improved Cheeger's inequality for finite graphs by Kwok et al. [27], which was subsequently used to establish an optimal dimension-free upper bound of eigenvalue ratios for weighted closed Riemannian manifolds with nonnegative Ricci curvature [33] (see also [34]). This answers open questions of Funano and Shioya [15, 16].

Spectral theory of discrete and continuous magnetic Laplacians attracted a lot of attention and literature on this subject developed rapidly, see, e.g., [9, 13, 14, 17, 24, 31, 41, 43, 47–49, 53]. Shigekawa proved the following comparison result in [47]: the least eigenvalue of the magnetic Laplacian on a closed Riemannian manifold is bounded from above by the least eigenvalue of a related Schrödinger operator. He also proved Weyl's asymptotic formula for magnetic Laplacians. Paternain [43] obtained an upper bound of the least eigenvalue in terms of the so-called harmonic value and Mañé's critical value of the corresponding Lagrangian. On finite planar graphs, Lieb and Loss [31] solved physically motivated extremality problems for eigenvalue expressions of the discrete magnetic Laplacian.

In this paper, we discuss a definition of Cheeger constants (Definitions 3.5, 3.6 and 7.3) reflecting the nontriviality of the magnetic potentials in terms of the frustration index (see Definitions 3.4 and 7.2) and the global connectivity of the underlying space. This definition works for both discrete and continuous magnetic Laplacians, and graph Laplacians with k-cyclic signatures (k∈Nk∈N). Recall that discrete magnetic Laplacians can be considered as graph Laplacians with a $U(1)$-signature. We would like to point out that our definition of Cheeger constants provides invariances under switching operations (Definition 2.3) or gauge transformations (Eq. 7.8). Furthermore, we prove the corresponding Cheeger inequalities and higher order Cheeger inequalities (Theorems 4.1, 4.6, 5.1, 7.4, and 7.7). We notice that our Theorem 4.6, the Cheeger inequality for discrete magnetic Laplacian, overlaps with a Cheeger inequality of Bandeira, Singer and Spielman [4, Theorem 4.1] in the framework of graph connection Laplacians [51]. See Remark 4.9 for a more detailed explanation. It is known in physics that "a magnetic field raises the energy" [31]. Roughly speaking, our estimates tell us that a magnetic field raises the energy via raising the frustration index. We focus on finite graphs and compact Riemannian manifolds in this paper.

Cheeger inequalities are essentially coarea inequalities. In the proof, we obtain in particular coarea inequalities related to the frustration index on graphs as well as on manifolds (Lemmas 4.3 and 7.5).

In fact, we were led to our Cheeger constant definition by an investigation of graph Laplacians with *k*-cyclic signatures, aiming at extending a previous spectral interpretation [3] of Harary's structural balance theory [21, 22] for graphs with (±1)(±1)-signatures. It turns out that the Cheeger inequalities for graph Laplacians with*k*-cyclic signatures and their proofs provide spectral clustering algorithms for partially oriented graphs (alternatively called mixed graphs without loops and multiple edges [23, 45, 46, 60]), aiming at detecting interesting substructures. A partially oriented graph may contain both oriented and unoriented edges. In the proof of such inequalities, we develop a random *k*-partition argument, which is algorithmic (see Lemma 4.2and Proposition 6.6). Recall that, in the setting of (±1)(±1)-signed graphs (i.e., k=2k=2), the eigenfunctions are real valued and a bipartition of the underlying graph can be given naturally according to the sign of the eigenfunction. But here we have complex valued eigenfunctions. Hence we do not have any natural *k*-partitions. That is why new ideas are needed. The generally non-symmetric graph Laplacians of partially oriented graphs are hardly useful for the purpose of spectral clustering. Our idea is to associate to a partially oriented graph and a natural number k∈Nk∈N an unoriented graph with a special *k*-cyclic signature. We then perform spectral clustering algorithms employing eigenfunctions of the graph Laplacian with the associated signature. According to our Cheeger constant definition, we can obtain interesting *k*-cyclic substructures. See Sect. 6 for details.

To prove higher order Cheeger inequalities, we develop new multi-way spectral clustering algorithms using metrics on *lens spaces* and *complex projective spaces*. This provides a deeper understanding of earlier spectral clustering algorithms via metrics on real projective spaces presented in [32] and [3]. These clustering algorithms were initially designed to find almost bipartite subgraphs of a given graph [32], and then extended to find almost balanced subgraphs of a signed graph [3]. While all operators studied in [3, 32] are bounded, we show that finding proper metrics for clustering is also useful for unbounded operators: the spectral clustering algorithms via metrics on complex projective spaces are crucial to prove the higher order Cheeger inequalities of the magnetic Laplacian on a closed Riemannian manifold (Lemma 7.8).

The paper is organized as follows. In Sect. 2, we set up notation for the discrete setting and recall basic spectral theory of related graph operators. In Sect. 3, we define the frustration index and the (multi-way) Cheeger constants. We prove the corresponding Cheeger's inequality in Sect. 4 and higher order Cheeger inequalities in Sect. 5. In Sect. 6, we discuss applications of Cheeger inequalities for spectral clustering on partially oriented graphs. In Sect. 7, we extend the results developed on discrete graphs to magnetic Laplacians on closed Riemannian manifolds.

2 NOTATIONS AND BASIC SPECTRAL THEORY

Throughout the paper, G=(V,E) denotes an undirected simple finite graph on *N* vertices with vertex set *V* and edge set *E*. We denote edges of *G* by {u,v}{u,v}, and u∼vu∼v means that u∈V and v∈V are connected by an edge. For any subset V˜⊆V, let G˜=(V˜,E˜) be the subgraph of *G* induced by V˜ that is, an edge {u,v} of G˜is an edge of *G* with u,v∈V˜u, We tacitly associate to every edge e={u,v}∈Ee={u,v}∈Ea positive symmetric weight wuv=wvu=

wewuv=wvu=we and define the weighted degree dudu of a vertex u∈V by du:=∑v,v∼uwuv For a positive measure μ:V→R+μ:, we define the *maximal* μμ-*degree* of the graph *G* as

$$d_{\mu} := \max_{u \in V} \left\{ \frac{\sum_{v,v\sim u} w_{uv}}{\mu(u)} \right\} = \max_{u \in V} \left\{ \frac{d_u}{\mu(u)} \right\}. \tag{2.1}$$

Henceforth we always consider weighted graphs, unless stated otherwise, but refer to them simply as graphs. We denote by e=(u,v) the oriented edge starting at *u* and terminating at *v*, and by e¯=(v,u) the oriented edge with the reversed orientation. Let E^{or}:={(u,v),(v,u)|{u,v}∈E} be the set of all oriented edges.

DEFINITION 2.1

Let *G* be a graph and ΓΓ be a group. A *signature* of *G* is a map s:Eor→Γ such that

$$s(\bar{e}) = s(e)^{-1}, \tag{2.2}$$

where s(e)−1 is the inverse of *s*(*e*) in Γ. The *trivial signature* s≡1, where 1 stands for the identity element of Γ, is denoted by s1. For an oriented edge e=(u,v)∈Eor, we will also write suv:=s(e) for convenience.

For k∈Nk∈N, we use the standard combinatorial notation [k]={1,2,...,k}In this paper, we will restrict ourselves to the case that the signature group ΓΓ is the cyclic group S1k:={ξj|j∈[k]} of order *k*, generated by the primitive *k*-th root of unity ξ:=e2πi/k∈C and the case that ΓΓ is the unitary group U(1)={z∈C||z|=1} The notation S1kSk1 emphasizes the fact that the elements in S1kSk1 lie on the unit circle.

We consider the following Laplacian ΔsμΔμs associated to the weighted graph (*G*, *w*) with signature s:Eor→Γs: vertex measure μ:V→R+μ: For any function f:V→Cf: and any vertex u∈V, we have

$$\Delta^s_\mu f(u) := \frac{1}{\mu(u)} \sum_{v, v\sim u} w_{uv}(f(u) - s_{uv} f(v)). \tag{2.3}$$

Note that the summation in (2.3) over the vertices *v* adjacent to *u* can also be understood as a summation over the oriented edges e=(u,v)∈Eor and the signature is evaluated at (*u*, *v*).

The Laplacian Δ^s_μ has the following decomposition

$$\Delta^s_\mu = (D_\mu)^{-1}(D - A^s)$$

where *D* and Dμ are the diagonal matrices with Duu=du and (Dμ)uu=μ(u) for all u∈V while AsAs is the (weighted) signed adjacency matrix with

$$A^s_{uv} := \begin{cases} 0, & u = v \text{ or } \{u, v\} \notin \text{ E}, \\ w_{uv}s_{uv}, & \{u, v\} \in \text{ E}. \end{cases}$$

When Γ=S1k, we call this operator the graph Laplacian with the *k*-cyclic signature. When Γ=U(1) this is the discrete magnetic Laplacian studied in Sunada [53] (see also Shubin [48]). By (2.2), the matrix ΔsμΔμs is Hermitian, and hence all its eigenvalues are real which can be listed with multiplicity as follows:

$$0 \le \lambda_1(\Delta^s_\mu) \le \lambda_2(\Delta^s_\mu) \le \cdots \le \lambda_N(\Delta^s_\mu) \le 2d_\mu. \tag{2.4}$$

For any two functions f,g:V→C, we define their inner product as

$$\langle f, g\rangle_\mu := \sum_{u\in V} f(u)\overline{g(u)}\mu(u). \tag{2.5}$$

It is easy to check that

$$\langle \Delta_\mu^s f, g\rangle_\mu = \sum_{\{u,v\}\in E} w_{uv}(f(u) - s_{uv} f(v))\overline{(g(u) - s_{uv} g(v))}. \quad (2.6)$$

Note that the right hand side of the above equality is well-defined since Γ⊆U(1) The corresponding*Rayleigh quotient* $\mathcal{R}_\mu^s(f)$ of a function f:V→C is

$$\mathcal{R}_\mu^s(f) := \frac{\sum_{\{u,v\}\in E} w_{uv}|f(u) - s_{uv} f(v)|^2}{\sum_{u\in V} |f(u)|^2 \mu(u)}. \quad (2.7)$$

The Courant–Fisher–Weyl min–max principle tells that, for any n∈[N]n∈[N],

$$\lambda_n(\Delta_\mu^s) = \min_{\substack{\langle f_p, f_q\rangle_\mu = 0,\\ p,q\in[n], p\neq q}} \max_{f\in \mathrm{span}\{f_1,\ldots,f_n\}} \mathcal{R}_\mu^s(f), \quad (2.8)$$

where $f_1, \ldots, f_n, f \not\equiv 0.$

Remark 2.2

In the case of a graph *G* with measure μd(u):=du for all u∈V and signature group Γ=U(1) or Γ=S1k, *k* even, Eq. (2.8) implies the following relations between eigenvalues

$$2 - \lambda_{N-k+1}(\Delta_{\mu_d}^{-s}) = \lambda_k(\Delta_{\mu_d}^s). \quad (2.9)$$

Here −s−s is the signature obtained by taking the negative values of *s* (as complex numbers). This generalizes [3, Lemma 1] where $\Gamma = S_2^1 = \{\pm 1\}$.

There is a natural operation, called switching, acting on the signatures [58, 59].

DEFINITION 2.3

Let *G* be a graph with signature *s*. For any function τ:V→Γτ:V→Γ we can define a new signature sτ:Eor→Γsτ:Eor→Γas follows:

$$s^{\tau}(e) = \tau(u)s(e)\tau(v)^{-1} \quad \forall e = (u, v) \in E^{or}. \tag{2.10}$$

We call the function ττ a *switching function*. The signature *s* and s′s′ are said to be *switching equivalent* if there exists a switching function ττ such that s′=sτ.

One can check that switching is indeed an equivalence relation on the set of signatures. An important invariant of the switching operation is the spectrum of Δ^{s}_{μ}. In fact, it holds that (see e.g. [59])

$$\Delta^{s^{\tau}}_{\mu} = D(\tau)\Delta^{s}_{\mu}D(\tau)^{-1}, \tag{2.11}$$

where D(τ) is the diagonal matrix with entries $D(\tau)_{uu}=\tau(u)$. This means that Δsτμ and ΔsμΔμs are unitarily equivalent and have the same spectrum. If the signature s:Eor→Γs is switching equivalent to the trivial signature s1s1, the operator Δ^{s}_{μ} is unitarily equivalent to the classical graph Laplacian. In this case we have λ1(Δ^{s}_{μ})=0 We will show in Sect. 4 that this is the only case where the first eigenvalue vanishes. Observe that on a tree, any signature is switching equivalent to the trivial signature.

Remark 2.4

The concept of switching is developed in the study of Harary's balance theory for signed graphs [21], i.e. graphs with signatures s: $E^{or} \rightarrow S_2^1 = \{+1, -1\}$ which we briefly review in the next section. The corresponding terminology in the magnetic theory is the gauge transformation, see, e.g., [9, 47]. Note that switching is an operation acting on the signatures suv:=eiαuvsuv:=eiαuv, while the gauge transformation is acting on the magnetic potentials αuvαuv, where (u,v)∈Eor. We will only use the terminology of the magnetic theory in the manifold case, see Sect. 7. Switching equivalent signatures are called cohomologous weight functions in [53].

3 FRUSTRATION INDEX AND CHEEGER CONSTANTS

One of our motivations for introducing the Cheeger constants is Harary's structural balance theory [21]. Let*G* be a finite graph with (possibly non-abelian) signature group Γ and signature s:Eor→Γ, and CC be a cycle, which is a graph of the sequence (u1,u2),(u2,u3),...,(ul−1,ul),(ul,u1) of distinct edges. Then the signature of C is the conjugacy class of the element

$$s_{u_1 u_2} s_{u_2, u_3} \cdots s_{u_{l-1} u_l} s_{u_l u_1} \in \Gamma.$$

Note that the signature of a cycle is switching invariant.

DEFINITION 3.1

A signature s:Eor→Γs is said to be *balanced* if the signature of every cycle of *G* is (the conjugacy class of the) identity element 1∈Γ

For convenience, we will also say that the graph *G* or a subgraph of *G* is balanced if the signature restricted on it is balanced. Since the signature of a cycle is switching invariant, the property of being balanced is also switching invariant. We have the following characterization of being balanced using switching operations.

PROPOSITION 3.2

([58], Corollary 3.3]) A signature s:Eor→Γ is balanced if and only if it is switching equivalent to the trivial signature s1s1.

Remark 3.3

The concept of balance has been studied in the literature under various terminologies. For example, a balanced cycle is said to be satisfying Kirchhoff's Voltage Law in [19]. In [9], the related concept to the signature of a cycle is the holonomy map. In magnetic theory, it is related to the magnetic flux [31].

We define the following frustration index to quantify how far a signature on a subset is from being balanced.

DEFINITION 3.4

Let *G* be a finite graph with signature *s* and V1⊆V nonempty with induced subgraph (V1,E1). The*frustration index* ιs(V1) of V1 is defined as

$$\iota^{s}(V_1) := \min_{\tau:V_1\to\Gamma} \sum_{\{u,v\}\in E_1} w_{uv}|\tau(u) - s_{uv}\tau(v)| \tag{3.1}$$

$$= \min_{\tau: V_1 \to \Gamma} \sum_{\{u,v\} \in E_1} w_{uv} |1 - \tau(u)^{-1} s_{uv} \tau(v)|. \tag{3.2}$$

A direct computation shows that the frustration index of a set is switching invariant and, according to Proposition 3.2, we have

ιs(V1)=0⇔ the subgraph induced by V1 is balanced. (3.3)

If *G* is unweighted and Γ={+1,−1} then

$$\iota^s(V) = 2e^s_{min}(V), \tag{3.4}$$

where $e^s_{min}(V)$ is the minimal number of edges that need to be removed from *E* in order to make G=(V,E) balanced. The quantity $e^s_{min}(V)$ is exactly the *line index of balance* of Harary [22]. Having the work of Vannimenus and Toulouse [56] in mind, Zaslavsky suggested later the term "frustration index" to Harary (T. Zaslavsky 2014, private communication).

We denote the *boundary measure* of V1V1 by

$$|E(V_1, V_1^c)| := \sum_{u \in V_1} \sum_{v \in V_1^c} w_{uv}, \tag{3.5}$$

where V_1^c is the complement of V1in *V*. The μμ-*volume* of V1is given by

$$\mathrm{vol}_\mu(V_1) := \sum_{u \in V_1} \mu(u). \tag{3.6}$$

DEFINITION 3.5

Let *G* be a finite graph with a signature *s*. The *Cheeger constant* $h^s_1(\mu)$is defined as

$$h_1^s(\mu) := \min_{\emptyset \neq V_1 \subseteq V} \phi_\mu^s(V_1), \tag{3.7}$$

where

$$\phi_\mu^s(V_1) := \frac{\iota^s(V_1) + |E(V_1, V_1^c)|}{\mathrm{vol}_\mu(V_1)}. \tag{3.8}$$

The choice of V achieving the minimum in (3.7) can be viewed as a subset of vertices which balances the two complementary goals of minimizing its frustration index and its expansion, measured by the edges $E(V_1,Vc_1)$ connecting V_1 with its complement.

A *nontrivial n-subpartition* of *V* is given by *n* pairwise disjoint nonempty subsets V1,...,Vn⊂V and a*nontrivial n-partition* additionally satisfies $\bigcup_{p\in[n]} V_p = V$.We abbreviate a nontrivial *n*-(sub)partition {V1,...,Vn} by {Vp}[n]. In the spirit of Miclo [39], we define the multi-way Cheeger constants as follows.

DEFINITION 3.6

Let *G* be a finite graph with a signature *s*. The *n-way Cheeger constant* hsn(μ)hns(μ) of *G* is defined as

$$h_n^s(\mu) := \min_{\{V_p\}_{[n]}} \max_{p\in[n]} \phi_\mu^s(V_p), \tag{3.9}$$

where the minimum is taken over all nontrivial *n*-subpartitions $\{V_p\}_{[n]}$ of V.

Observe that the *n*-way Cheeger constant of a graph *G* is monotone with respect to *n*, that is, $h_n^s(\mu) \leq h_{n+1}^s(\mu)$.

Using (3.3) and the fact that the frustration index is switching invariant, we obtain the following properties of the Cheeger constants.

PROPOSITION 3.7

The n-way Cheeger constants $h_n^s(\mu)$ of a graph G are switching invariant. Moreover, $h_n^s(\mu) = 0$ if and only if G consists of at least n connected components and at least n of them are balanced.

If sb:Eor→Γ denotes a balanced signature, then $h_n^s(\mu) = 0$ becomes trivial and

$$h_2^{Sb}(\mu) = \min_{\{V_1,V_2\}} \max_{p\in[2]} \phi_\mu^{Sb}(V_p) = \min_{\substack{\emptyset\neq V_1\subseteq V \\ \mathrm{vol}_\mu(V_1)\leq\frac{1}{2}\mathrm{vol}_\mu(V)}} \frac{|E(V_1, V_1^c)|}{\mathrm{vol}_\mu(V_1)}, \qquad (3.10)$$

that is, $h_n^s(\mu)$ reduces to the classical Cheeger constant.

Remark 3.8

Due to equation (3.4), the *n*-way Cheeger constant in (3.9) reduces to the signed Cheeger constant introduced on signed graphs [3] with signature group Γ={+1,−1} We mention that the signed Cheeger constant in [3] is a unification of the classical Cheeger constant, the non-bipartiteness parameter in [10], the bipartiteness ratio in [54, 55], and the dual Cheeger constant in [5].

For n∈[N] and any signature s:Eor→Γ, we observe

$$h_n^{Sb}(\mu) \leq h_n^s(\mu). \qquad (3.11)$$

In fact, if $h_n^s(\mu) = \max_{p\in[n]} \phi_\mu^s(\widetilde{V}_p)$, i.e. $\{\widetilde{V}_p\}_{[n]}$ is the nontrivial n subpartition of V that achieves $h_n^s(\mu)$, we have $\phi_\mu^{S_b}(\widetilde{V}_p) \leq \phi_\mu^s(\widetilde{V}_p)$ since ιsb $\iota^{S_b}(\widetilde{V}_p) = 0 \leq \iota^{S_b}(\widetilde{V}_p)$. Hence, (3.11) follows by Definition3.6. The inequality (3.11) is similar, in spirit, with Kato's inequality for noncompact spaces [12, Lemma 1.2, Corollary 1.3] (alternatively, also called the diamagnetic inequality for both compact and noncompact spaces in [31]) where the bottom of the spectrum increases when a balanced signature is replaced by an unbalanced signature.

For n=1we have the following result. Recalling $h_n^s(\mu) = 0$,Proposition 3.9 tells us that this change of the first Cheeger constant (by choosing an unbalanced signature) can be quite large.

PROPOSITION 3.9

Let G be an unweighted connected finite d-regular graph and M=$\max_{v\in V} \mu(v)$Then, for every k≥2, there exists a k-cyclic signature so: $E^{or} \to S_k^1$ such that

$$h_1^{so}(\mu) \geq \frac{d - 2\sqrt{d-1}}{2M}. \tag{3.12}$$

Proof

Extending a result of [37, 38], it is shown in [35, Theorem 2] that there exists a k-cyclic signature s0s0 such that the maximal eigenvalue of the matrix As0As0 is no greater than $2\sqrt{d-1}$. The estimate (3.12) is then an immediate consequence of this result, combined with Cheeger's inequality (4.1), given at the beginning of the next section.

4 CHEEGER'S INEQUALITY

In this section, we prove Cheeger's inequality relating $\lambda_1(\Delta^s_\mu)$ to the first Cheeger constant $h^s_1(\mu)$ for graph Laplacians with cyclic signatures (Theorem 4.1) and for discrete magnetic Laplacians (Theorem 4.6).

THEOREM 4.1

Let *G* be a finite graph with signature s: $E^{or} \to S^1_k$. Then we have

$$\frac{1}{2}\lambda_1(\Delta^s_\mu) \leq h^s_1(\mu) \leq 2\sqrt{2d_\mu\lambda_1(\Delta^s_\mu)}. \tag{4.1}$$

We start with preparations for the proof of Theorem 4.1. Let $B_r(0) := \{z \in \mathbb{C} \mid |z| < r\}$ be the open disk in C with center 0 and radius *r*. For $\theta \in [0, 2\pi)$ and $k \in \mathbb{N}$, we define the following *k* disjoint sectorial regions

$$Q^\theta_j := \left\{ re^{i\alpha} \in \overline{B_1(0)} \;\middle|\; r \in (0,1], \alpha \in \left[\theta + \frac{2\pi j}{k}, \theta + \frac{2\pi(j+1)}{k}\right)\right\}, \tag{4.2}$$

where j=0,1,...,k−1. Then for any t∈[0,1], we define the function $Y_{t,\theta} : \overline{B_1(0)} \to \mathbb{C}$ as

$$Y_{t,\theta}(z) := \begin{cases} \xi^j, & \text{if } z \in Q^\theta_j \setminus B_t(0), \\ 0, & \text{if } z \in B_t(0), \end{cases} \tag{4.3}$$

where ξ denotes the *k*-th primitve root of unity.

The following lemma plays a key role.

LEMMA 4.2

For any two points $z_1, z_2 \in \overline{B_1(0)}$, we have

$$\frac{1}{2\pi}\int_0^{2\pi}\int_0^1 \left|Y_{\sqrt{t},\theta}(z_1) - Y_{\sqrt{t},\theta}(z_2)\right| dt\, d\theta \le 2|z_1 - z_2|\,(|z_1| + |z_2|). \tag{4.4}$$

Proof

W.l.o.g., we can assume that $|z_1| \ge |z_2|$ with $z_1 \in Q^\theta_{j_1}$ and $z_2 \in Q^\theta_{j_2}$. Then we have

$$|Y_{\sqrt{t},\theta}(z_1) - Y_{\sqrt{t},\theta}(z_2)| = \begin{cases} |\xi^{j_1} - \xi^{j_2}|, & \text{if } \sqrt{t} \le |z_2|, \\ 1, & \text{if } |z_2| < \sqrt{t} \le |z_1|, \\ 0, & \text{if } |z_1| < \sqrt{t}. \end{cases} \tag{4.5}$$

Hence,

$$\int_0^1 \left|Y_{\sqrt{t},\theta}(z_1) - Y_{\sqrt{t},\theta}(z_2)\right| dt = |\xi^{j_1} - \xi^{j_2}| \cdot |z_2|^2 + (|z_1|^2 - |z_2|^2). \tag{4.6}$$

Let $\alpha_{z_1z_2} \in [0, \pi]$ be the angle between the two rays joining z1,z2 to the origin. If $2\pi l/k \le \alpha_{z_1z_2} < 2\pi(l+1)/k$ for some integer 0≤l<k/, the term $|\xi^{j_1} - \xi^{j_2}|$ is equal to either $|1 - \xi^l|$ or $|1 - \xi^{l+1}|$, hence we calculate

$$\frac{1}{2\pi}\int_0^{2\pi}\int_0^1 \left|Y_{\sqrt{t},\theta}(z_1) - Y_{\sqrt{t},\theta}(z_2)\right| dt\, d\theta$$

$$\begin{aligned}
&= \left(\frac{k\alpha_{z_1 z_2}}{2\pi} - l\right)\left(|1 - \xi^{l+1}| \cdot |z_2|^2 + |z_1|^2 - |z_2|^2\right) \\
&\quad + \left(l + 1 - \frac{k\alpha_{z_1 z_2}}{2\pi}\right)\left(|1 - \xi^l| \cdot |z_2|^2 + |z_1|^2 - |z_2|^2\right) \\
&\le 2|1 - \xi^l| \cdot |z_2|^2 + \left(|z_1|^2 - |z_2|^2\right),
\end{aligned}$$

where we used $|1 - \xi^{l+1}| \le |1 - \xi| + |1 - \xi^l| \le 2|1 - \xi^l|$. Observe that we have

$$|z_1 - z_2| \ge \left|\frac{z_1}{|z_1|}|z_2| - z_2\right| \ge |z_2| \cdot |1 - \xi^l| \tag{4.7}$$

and

$$|z_1|^2 - |z_2|^2 = (|z_1| - |z_2|) \cdot (|z_1| + |z_2|) \le |z_1 - z_2| \cdot (|z_1| + |z_2|). \tag{4.8}$$

Therefore, we obtain

$$\frac{1}{2\pi}\int_0^{2\pi}\int_0^1 \left|Y_{\sqrt{t},\theta}(z_1) - Y_{\sqrt{t},\theta}(z_2)\right| dt\, d\theta \le 2|z_1 - z_2| \cdot |z_2| + |z_1 - z_2| \cdot (|z_1| + |z_2|). \tag{4.9}$$

which implies (4.4).

Lemma 4.2 can be considered as an extension of [3, Lemma 5] and [54, 55, Section 3.2]. The novel point here is that we introduce an extra degree of randomness in the argument of *z* in order to handle the difficulty caused by cyclic signatures. Actually, this provides a random k-partition parametrized by an angle θθ, which will be discussed further in Sect. 6. This lemma is a version of a coarea inequality, which becomes transparent from the following direct consequence.

For any non-zero function f:V→C defined on the vertices of a graph G and any $t \in [0, \max_{u \in V} |f(u)|]$, we define the following non-empty subset of V:

$$V^f(t) := \{u \in V \mid t \le |f(u)|\}. \tag{4.10}$$

LEMMA 4.3

(Coarea inequality) Let $E^{or} \to S_k^1$ be a signature of G. For any function f:V→C with $\max_{u \in V} |f(u)| = 1$, we have

$$\begin{aligned} &\int_0^1 \iota^s\left(V^f(\sqrt{t})\right) + \left|E\left(V^f(\sqrt{t}), (V^f(\sqrt{t}))^c\right)\right| dt \\ &\le 2 \sum_{\{u,v\} \in E} w_{uv} \left|f(u) - s_{uv} f(v)\right| \cdot (|f(u)| + |f(v)|). \end{aligned} \tag{4.11}$$

Proof

First observe that

$$\begin{aligned} &\frac{1}{2\pi} \int_0^{2\pi} \int_0^1 \sum_{\{u,v\} \in E} w_{uv} \left|Y_{\sqrt{t},\theta}(f(u)) - s_{uv} Y_{\sqrt{t},\theta}(f(v))\right| dt\, d\theta \\ &\ge \int_0^1 \iota^s\left(V^f(\sqrt{t})\right) + \left|E(V^f(\sqrt{t}), (V^f(\sqrt{t}))^c)\right| dt. \end{aligned} \tag{4.12}$$

In fact, the summation in the integrand of the LHS of the above inequality can be split into two parts: (i) The summation over edges connecting two vertices from $V^f(\sqrt{t})$ and $V^f(\sqrt{t})^c$,respectively. This part equals to $|E(V^f(\sqrt{t}),(V^f(\sqrt{t}))^c)|$; (ii) The summation over edges connecting two vertices from $V^f(\sqrt{t})$This part is bounded from below by $\iota^s\,(V^f(\sqrt{t}))$by Definition 3.4.

Notice further that

$$s_{uv}Y_{\sqrt{t},\theta}(f(v)) = Y_{\sqrt{t},\theta}(s_{uv}f(v)), \tag{4.13}$$

the inequality (4.11) follows directly from Lemma 4.2.

The coarea inequality is particularly useful to prove Lemma 4.4.

LEMMA 4.4

Let $s : E^{or} \to S_k^1$be a signature of *G* and f:V→Cf:V→C be a nonzero function. Then there exists $\in [0, \max_{u\in V} |f(u)|^2]$such that

$$\phi_\mu^s(V^f(\sqrt{t'})) \leq 2\sqrt{2d_\mu \mathcal{R}_\mu^s(f)}, \tag{4.14}$$

where $\mathcal{R}_\mu^s(f)$was defined in (2.7).

Proof

Since *f* is non-zero, we may assume (after rescaling) that $\max_{u\in V} |f(u)| = 1$. Moreover,

$$|Y_{\sqrt{t},\theta}(f(u))| = \begin{cases} 1, & \text{if } |f(u)| \geq \sqrt{t}, \\ 0, & \text{otherwise}, \end{cases} \tag{4.15}$$

implies

$$\int_0^1 \mathrm{vol}_\mu(V^f(\sqrt{t}))\,dt = \int_0^1 \sum_{u\in V} \left|Y_{\sqrt{t},\theta}(f(u))\right| \mu(u)\,dt = \sum_{u\in V} |f(u)|^2 \mu(u). \tag{4.16}$$

Now we consider the quotient

$$I := \frac{\int_0^1 \iota^s(V^f(\sqrt{t})) + \left|E(V^f(\sqrt{t}), (V^f(\sqrt{t}))^c)\right|\,dt}{\int_0^1 \mathrm{vol}_\mu(V^f(\sqrt{t}))dt}. \tag{4.17}$$

Therefore, there exists t′∈[0,1] such that

$$I \geq \phi_\mu^s(V^f(\sqrt{t'})). \tag{4.18}$$

On the other hand, Lemma 4.3, (4.16), and the Cauchy–Schwarz inequality imply

$$\begin{aligned} I &\leq \frac{2\sum_{\{u,v\}\in E} w_{uv}|f(u) - s_{uv}f(v)| \cdot (|f(u)| + |f(v)|)}{\sum_{u\in V} |f(u)|^2 \mu(u)} \\ &\leq \frac{2\sqrt{\sum_{\{u,v\}\in E} w_{uv}|f(u) - s_{uv}f(v)|^2}\sqrt{\sum_{\{u,v\}\in E} w_{uv}(|f(u)| + |f(v)|)^2}}{\sum_{u\in V} |f(u)|^2 \mu(u)}. \end{aligned}$$

Since

$$\begin{aligned} \sum_{\{u,v\}\in E} w_{uv}\,(|f(u)| + |f(v)|)^2 &\leq 2\sum_{\{u,v\}\in E} w_{uv}(|f(u)|^2 + |f(v)|^2) \\ &= 2\sum_{u\in V}\sum_{v,v\sim u} w_{uv}|f(u)|^2, \end{aligned}$$

we conclude that

$$I \leq 2\sqrt{2d_\mu \mathcal{R}^s_\mu(f)}. \tag{4.19}$$

Combining the estimates (4.18) and (4.19) proves the lemma.

Proof of Theorem 4.1

The upper estimate in (4.1) follows from Lemma 4.4 by setting *f* to be the eigenfunction corresponding to the eigenvalue λ1(Δsμ)λ1(Δμs).

It remains to prove the lower estimate of hs1(μ)h1s(μ) in (4.1). Let V˜be the subset of *V* that achieves the Cheeger constant hs1(μ)h1s(μ) in (3.7) with induced subgraph (V˜,E˜) and τ˜:V˜→S1k be the switching function that achieves the frustration index ιs(V˜) in (3.1). Define the function f˜:V→C via:

$$\widetilde{f}(u) := \begin{cases} \widetilde{\tau}(u), & \text{if } u \in \widetilde{V}, \\ 0, & \text{otherwise}. \end{cases} \tag{4.20}$$

Using (2.8) and the estimate $|\widetilde{\tau}(u) - s_{uv}\widetilde{\tau}(v)|$, we obtain

$$\begin{aligned} \lambda_1(\Delta^s_\mu) &\leq \mathcal{R}^s_\mu(\widetilde{f}) \\ &= \frac{\sum_{\{u,v\}\in\widetilde{E}} w_{uv}|\widetilde{\tau}(u) - s_{uv}\widetilde{\tau}(v)|^2 + |E(\widetilde{V},\widetilde{V}^c)|}{\mathrm{vol}_\mu(\widetilde{V})} \\ &\leq \frac{2\iota^s(\widetilde{V}) + |E(\widetilde{V},\widetilde{V}^c)|}{\mathrm{vol}_\mu(\widetilde{V})} \\ &\leq 2h^s_1(\mu). \end{aligned} \tag{4.21}$$

Remark 4.5

Since the signature is S_k^1-valued, the constant 2 in (4.21) can be slightly improved to be $|\,|1-\xi^{(k-1)/2}|$ when k is odd.

For Γ=U(1) we have the following Cheeger's inequality.

THEOREM 4.6

Let G be a finite graph with signature $s : E^{or} \to U(1)$.Then

$$\frac{1}{2}\lambda_1(\Delta_\mu^s) \leq h_1^s(\mu) \leq \frac{3}{2}\sqrt{2d_\mu \lambda_1(\Delta_\mu^s)}. \tag{4.22}$$

The constant in the upper bound of (4.22) is slightly better than the constant in (4.1). This is due to Lemma4.7 below.

For any t∈[0,1], we define $X_t : \overline{B_1(0)} \to \mathbb{C}$ as

$$X_t(z) := \begin{cases} z/|z|, & \text{if } z \in \overline{B_1(0)} \setminus B_t(0), \\ 0, & \text{if } z \in B_t(0). \end{cases} \tag{4.23}$$

LEMMA 4.7

For any two points $z_1, z_2 \in \overline{B_1(0)}$,we have

$$\int_0^1 \left| X_{\sqrt{t}}(z_1) - X_{\sqrt{t}}(z_2) \right| dt \leq \frac{3}{2}|z_1 - z_2|(|z_1| + |z_2|). \tag{4.24}$$

Proof

W.l.o.g., we assume that |z1|≥|z2|>0. Observe that

$$\int_0^1 \left| X_{\sqrt{t}}(z_1) - X_{\sqrt{t}}(z_2) \right| dt \le \left| \frac{z_1}{|z_1|} - \frac{z_2}{|z_2|} \right| |z_2|^2 + (|z_1|^2 - |z_2|^2). \tag{4.25}$$

Recalling (4.7), we have

$$\left| \frac{z_1}{|z_1|} - \frac{z_2}{|z_2|} \right| |z_2|^2 \le |z_1 - z_2||z_2| \le \frac{1}{2}|z_1 - z_2|(|z_1| + |z_2|). \tag{4.26}$$

Combining this with (4.8) proves the lemma.

With this lemma at hand, the proofs of Theorems 4.1 and 4.6 are very similar. We omit the details but mention the following analogue of Lemma 4.4.

LEMMA 4.8

Let s:E^or→U(1)s) be a signature of *G* and f:V→C be a nonzero function. Then there exists $t' \in [0, \max_{u \in V} |f(u)|^2]$ such that

$$\phi_\mu^s(V^f(\sqrt{t'})) \le \frac{3}{2}\sqrt{2 d_\mu \mathcal{R}_\mu^s(f)}. \tag{4.27}$$

Remark 4.9

We notice that the inequality (4.22) for Γ=U(1) overlaps with a Cheeger inequality for a connection Laplacian of *G* discussed by

Bandeira, Singer and Spielman [4] to solve a partial synchronization problem. The connection Laplacian LL is defined for a simple graph *G* where a matrix $O_{uv} \in O(l)$ is assigned to each $(u, v) \in E^{or}$ such that $O_{vu} = (O_{uv})^{-1}$For any vector-valued function $f : V \to \mathbb{R}^l$ and any vertex $u \in V$, we then have

$$\mathcal{L}f(u) := \frac{1}{d_u} \sum_{v, v \sim u} w_{uv}(f(u) - O_{uv}f(v)) \in \mathbb{R}^l. \tag{4.28}$$

For a graph *G* with signature s: $E^{or} \to U(1)$we consider the particular positive measure μμ on *V* defined as μ(u):=duμ(u):=du and rewrite the value $s_{uv} := a_{uv} + ib_{uv} \in U(1)$for each $(u, v) \in E^{or}$ as

$$\begin{pmatrix} a_{uv} & -b_{uv} \\ b_{uv} & a_{uv} \end{pmatrix} \in SO(2). \tag{4.29}$$

If we also rewrite a complex valued function $f := f_1 + if_2$as an $\mathbb{R}^2$-valued function $f := (f_1, f_2)^T$, the discrete magnetic Laplacian Δ^s_μ translates into a connection Laplacian Ls with eigenvalues

$$0 \le \lambda_1(\Delta^s_\mu) = \lambda_1(\Delta^s_\mu) \le \cdots \le \lambda_N(\Delta^s_\mu) = \lambda_N(\Delta^s_\mu). \tag{4.30}$$

Thus, each eigenvalue $\lambda_i(\Delta^s_\mu)$ of Δ^s_μ an eigenvalue of LsLs with doubled multiplicity. If we denote the Euclidean norm in $\mathbb{R}^l$ by $\|\cdot\|$,Bandeira, Singer and Spielman define a *(partial ℓ1ℓ1-frustration constant* as

$$\eta^*_{G,1} := \min_{\tau : V \to \mathbb{S}^{l-1} \cup \{0\}} \frac{\sum_{u,v \in V} w_{uv} \|\tau(u) - O_{uv}\tau(v)\|}{\sum_{u \in V} d_u \|\tau(u)\|}, \tag{4.31}$$

and prove that

$$\lambda_1(\mathcal{L}) \leq \eta^*_{G,1} \leq \sqrt{10\lambda_1(\mathcal{L})}. \tag{4.32}$$

If we assign elements of SO(2) to edges of G (instead of $O(2)$), we observe that

$$\eta^*_{G,1} = 2h^s_1(\mu), \quad \text{and} \quad \lambda_1(\mathcal{L}^s) = \lambda_1(\Delta^s_\mu). \tag{4.33}$$

Hence, inequality (4.32) leads to inequality (4.22). Finally, Bandeira, Singer and Spielman have a refined analysis for (4.24) that improves the constant 3 / 2 in (4.24) and (4.22) to 5√/25/2, [4, Appendix A].

A direct corollary of Theorems 4.1 and 4.6 as well as Proposition 3.7 is the following characterization of the case that the first eigenvalue vanishes.

COROLLARY 4.10

$\lambda_1(\Delta^s_\mu) = 0$ if and only if the underlying graph has a balanced connected component.

We remark that Corollary 4.10 can also be easily derived by the min-max principle (2.8).

5 SPECTRAL CLUSTERING VIA LENS SPACES AND COMPLEX PROJECTIVE SPACES

In this section, we prove the following higher order Cheeger inequalities.

THEOREM 5.1

There exists an absolute constant C>0 such that for any finite graph *G* with signature *s* and all n∈[N], we have

$$\frac{1}{2}\lambda_n(\Delta_\mu^s) \le h_n^s(\mu) \le Cn^3\sqrt{d_\mu \lambda_n(\Delta_\mu^s)}. \tag{5.1}$$

Note that in Theorem 5.1 the signature group ΓΓ can be either S1kSk1 or *U*(1).

The upper bound of $h_n^s(\mu)$ in (5.1) is the essential part of Theorem 5.1 and its proof relies on the development of a proper spectral clustering algorithm for the operator Δ_μ^s. In other words, we aim to find an*n*-subpartition $\{V_p\}_{[n]}$ with small constants $\phi_\mu^s(V_p)$, based on the information contained in the eigenfunctions of the operator Δ_μ^s.

Let *fi* be an orthonormal family of eigenfunctions corresponding to $\lambda_i(\Delta_\mu^s)$ for $i \in [n]$. We consider the following map:

$$F : V \to \mathbb{C}^n, \quad F(u) = (f_1(u), f_2(u), \ldots, f_n(u)). \tag{5.2}$$

Since $\lambda_n(\Delta_\mu^s) = \mathcal{R}_\mu^s(f_n)$, the Rayleigh quotient of *F* is also bounded by $\lambda_n(\Delta_\mu^s)$:

$$\begin{aligned}\mathcal{R}_\mu^s(F) &:= \frac{\sum_{\{u,v\}\in E} w_{uv}\|F(u) - s_{uv}F(v)\|^2}{\sum_{u\in V}\mu(u)\|F(u)\|^2} \\ &= \frac{\sum_{p\in[n]}\sum_{\{u,v\}\in E} w_{uv}|f_p(u) - s_{uv}f_p(v)|^2}{\sum_{p\in[n]}\sum_{u\in V}\mu(u)|f_p(u)|^2} \\ &\le \lambda_n(\Delta_\mu^s),\end{aligned} \tag{5.3}$$

where ‖·‖‖·‖ stands for the standard Hermitian norm in CnCn. Our goal is to construct *n* maps $\Psi_p : V \to \mathbb{C}^n$, $p \in [n]$,with pairwise disjoint supports such that

1 each Ψp can be viewed as a localization of *F*, i.e., ΨpΨp is the product of *F* and a cut-off function η:V→Rη (see 5.13 below),

2 each Rayleigh quotient satisfies Rsμ(Ψp)≤C(n)Rsμ(F), where *C*(*n*) is a constant only depending on *n*.

Then, applying Lemmas 4.4 and 4.8 will finish the proof.

This strategy is adapted from the proof of the higher order Cheeger inequalities for unsigned graphs due to Lee et al. [28, 29]. A critical new point here is to find a proper metric on the space of points $\{F(u)|u \in V\} \subset \mathbb{C}^n$ Cn for the spectral clustering algorithm. In other words, we need a proper metric to localize the map *F*. The original algorithm in [28, 29] used a spherical metric. The second author [32] studied a spectral clustering via metrics on real projective spaces to prove higher order dual Cheeger inequalities for unsigned graphs. Later in [3], the above two algorithms and, hence, the corresponding two kinds of inequalities, were unified in the framework of Harary's signed graphs, i.e., graphs with signatures s:Eor→{+1,−1} In particular, the metrics on real projective spaces were shown to be the proper metrics for clustering in the framework of signed graphs. In our current more general setting of graphs with signatures s:Eor→Γs:Eor→Γ, where Γ=S1k or Γ=U(1) the new metrics will be defined on lens spaces and complex projective spaces.

5.1 Lens spaces and complex projective spaces

In this subsection, we provide metrics of lens spaces and complex projective spaces for the spectral clustering algorithms in the case of Γ=S1k and Γ=U(1) respectively. Both lens spaces and complex projective spaces are important objects in geometry and topology. See, e.g., [26, Chapter 5] for details about these spaces.

Let $\mathbb{S}^{2n-1} := \{\mathbf{z} \in \mathbb{C}^n \mid \|\mathbf{z}\| = 1\}$ be the unit sphere in the space CnCn. Then Γ⊂C acts on S2n−1by scalar multiplication. For any two points $\mathbf{z}_1, \mathbf{z}_2 \in \mathbb{S}^{2n-1} \subset \mathbb{C}^n$, we define the following equivalence relation:

$$\mathbf{z}_1 \sim \mathbf{z}_2 \Leftrightarrow \exists \gamma \in \Gamma \quad \text{such that } \mathbf{z}_1 = \gamma \mathbf{z}_2. \tag{5.4}$$

For $\Gamma = S^1_k$, the corresponding quotient space $\mathbb{S}^{2n-1}/\Gamma$ is the lens space L(k;1,...,1) while for Γ=U(1), the quotient space $\mathbb{S}^{2n-1}/\Gamma$ is the complex projective space $\mathbb{C}P^{n-1}$.. Let [z] denote the equivalence class of $\mathbf{z} \in \mathbb{S}^{2n-1}$. We consider the following metric on S2n−1/ΓS2n−1/Γ:

$$d([\mathbf{z}_1], [\mathbf{z}_2]) := \min_{\gamma \in \Gamma} \|\mathbf{z}_1 - \gamma \mathbf{z}_2\|. \tag{5.5}$$

The space $\mathbb{S}^{2n-1}/\Gamma$ can also be endowed with a distance d_{quot} which is induced from the standard Riemannian metric on S2n−1⊂R2nS2n−1⊂R2n. This induced metric has positive Ricci curvature. If $\Gamma = S^1_k$, the sectional curvature of this metric is constant equal to 1, and if Γ=U(1)Γ=U(1), this metric is the well-known Fubini-Study metric. The two metrics *d* and dquotdquot on $\mathbb{S}^{2n-1}/\Gamma$ are equivalent, i.e., there exist two constantsc1,c2>0c1,c2>0 such that for all $[z_1], [z_2] \in S^{2n-1}/\Gamma$,

$$c_1 d_{quot}([\mathbf{z}_1], [\mathbf{z}_2]) \leq d([\mathbf{z}_1], [\mathbf{z}_2]) \leq c_2 d_{quot}([\mathbf{z}_1], [\mathbf{z}_2]). \tag{5.6}$$

Recall the concept of the metric doubling constant ρX of a metric space (X,dX). This constant is the infimum of all numbers ρρ such that every ball *B* in X can be covered by ρ balls of half the radius of *B*.

PROPOSITION 5.2

The metric doubling constant ρΓ of $(\mathbb{S}^{2n-1}/\Gamma, d)$ satisfies

$$\log_2 \rho_\Gamma \leq Cn, \tag{5.7}$$

where *C* is an absolute constant.

Proof

By equivalence (5.6), we only need to consider the metric space (S2n−1/Γ,dquot). Since $\mathbb{S}^{2n-1}/\Gamma$ with its standard metric has nonnegative Ricci curvature, the Bishop-Gromov comparison theorem guarantees

$$\frac{\mathrm{vol}(B_r([\mathbf{z}_1]))}{\mathrm{vol}(B_{r/2}([\mathbf{z}_1]))} \leq \bar{C}^n. \tag{5.8}$$

for some absolute constant C¯. (Note that the real dimension of the lens space is 2n−1 and of the complex projective space is 2n−2.) A standard argument implies now the claim of the proposition. For details see, e.g., [8, p. 67] or [32, Section 2.2].

The metric d on $\mathbb{S}^{2n-1}/\Gamma$ induces a pseudo metric on the space $\mathbb{C}^n\setminus\{0\}$, which—by abuse of notation—will again be denoted by d:

$$d(\mathbf{z}_1, \mathbf{z}_2) := d\left(\left[\frac{\mathbf{z}_1}{\|\mathbf{z}_1\|}\right], \left[\frac{\mathbf{z}_2}{\|\mathbf{z}_2\|}\right]\right). \tag{5.9}$$

The following obvious property is the reason why we use the metric d on $\mathbb{S}^{2n-1}/\Gamma$ from (5.5). This reason will become clear in the next Sect. 5.2.

PROPOSITION 5.3

For every pair $\mathbf{z}_1, \mathbf{z}_2 \in \mathbb{C}^n \setminus \{0\}$ and every $\gamma\in\Gamma$, we have

$$d(\mathbf{z}_1, \mathbf{z}_2) = d(\mathbf{z}_1, \gamma\mathbf{z}_2). \tag{5.10}$$

The considerations of the next two subsections prepare the ground for the study of the Rayleigh quotient $\mathcal{R}^s_\mu(F)$ of the map $F:V\to\mathbb{C}^n$ defined in (5.2).

5.2 Localization of the map *F*

We endow the support $V_F := \{u \in V\,|\,F(u) \neq 0\}$ with the pseudo metric d_F induced by d via

$$d_F(u, v) := d(F(u), F(v)). \tag{5.11}$$

Given a subset $S\subseteq V$ and $\epsilon>0$, we first define a cut-off function $\eta:V\to\mathbb{R}$ by

$$\eta(u) := \begin{cases} 0, & \text{if } F(u) = 0, \\ \max\{0, 1 - \frac{1}{\epsilon} d_F(u, S \cap V_F)\}, & \text{otherwise} \end{cases} \quad (5.12)$$

and then localize *F* via η as

$$\Psi := \eta F : V \to \mathbb{C}^n. \quad (5.13)$$

Note that the ϵϵ-neighborhood $N_\epsilon(S \cap V_F, d_F) := \{u \in V \mid d_F(u, S \cap V_F) < \epsilon\} < \epsilon\}$ of $S \cap V_F$ contains the support of the map ΨΨ.

In the next lemma, $G_F=(V_F,E_F)$ denotes the induced subgraph on VF of *G*.

LEMMA 5.4

$$\text{If } \{u, v\} \in E_F \text{ and } \|F(v)\| \le \|F(u)\| \text{ then} \quad (5.14)$$
$$d(F(u), F(v))\|F(v)\| \le \|F(u) - s_{uv} F(v)\|.$$

Proof

Observe that we only need to prove

$$d(F(u), F(v))\|F(v)\| \le \|F(u) - F(v)\| \quad (5.15)$$

for any pair of points $F(u), F(v) \in \mathbb{C}^n \setminus \{0\}$ with $\|F(v)\| \le \|F(u)\|$:we can replace *F*(*v*) in (5.15) by $s_{uv} F(v)$and use Proposition 5.3 to obtain (5.14). By the definition of the metric *d*, we obtain (5.15) as follows:

$$d(F(u), F(v))\|F(v)\| \le \left\| \frac{F(u)}{\|F(u)\|} - \frac{F(v)}{\|F(v)\|} \right\| \|F(v)\| \le \|F(u) - F(v)\|.$$

where we used the estimate (4.7) for the latter inequality.

Lemma 5.4 enables us to prove the following result.

LEMMA 5.5

For any {u,v}∈E, we have

$$\|\Psi(u) - s_{uv}\Psi(v)\| \leq \left(1 + \frac{1}{\epsilon}\right) \|F(u) - s_{uv}F(v)\|. \qquad (5.16)$$

Proof

If at least one of *F*(*u*) and *F*(*v*) is equal to zero, then the estimate (5.16) holds trivially. Hence, we suppose that u,v∈VFu,v∈VF. W.l.o.g., we can assume that ‖F(u)‖≤‖F(v)‖ and calculate

$$\begin{aligned}\|\Psi(u) - s_{uv}\Psi(v)\| &= \|\eta(u)F(u) - s_{uv}\eta(v)F(v)\| \\ &\leq |\eta(u)| \cdot \|F(u) - s_{uv}F(v)\| + |\eta(u) - \eta(v)| \cdot \|F(v)\| \\ &\leq \|F(u) - s_{uv}F(v)\| + \frac{d_F(u, v)\|F(v)\|}{\epsilon}.\end{aligned}$$

Applying Lemma 5.4 completes the proof. □

Note that the inequality (5.16) is useful for the estimate of the numerator of the Rayleigh quotient of Ψ.

5.3 Decomposition of the underlying space via orthonormal functions

For later purposes, we work on a general measure space (V,μ) in this subsection, where V is a topological space and μμ is a Borel measure. Two particular cases we have in mind are a vertex set *V* of a finite graph with a measure μ:V→R+ ,and a closed Riemannian manifold with its Riemannian volume measure. We will apply the results in this subsection to the latter case in Sect. 7.

On (V,μ), we further assume that there exist *n* measurable functions

$$f_1, f_2, \ldots, f_n : \mathcal{V} \to \mathbb{C},$$

which are orthonormal, i.e., for any i,j∈[n],

$$\langle f_i, f_j \rangle := \int_{\mathcal{V}} f_i \overline{f_j} d\mu = \delta_{ij}.$$

Then the map F:V→Cn is given accordingly as in (5.2).

We consider the measure μF on V given by

$$d\mu_F = \|F\|^2 d\mu.$$

For any two points x, y in $\mathcal{V}_F := \{x \in \mathcal{V} : F(x) \neq 0\}$, we have the distance between them

$$d_F(x, y) := \min_{\gamma \in \Gamma} \left\| \frac{F(x)}{\|F(x)\|} - \gamma \frac{F(y)}{\|F(y)\|} \right\|. \tag{5.17}$$

The main result of this subsection is the following theorem.

THEOREM 5.6

Let (VF,dF,μF) be as above. There exist an absolute constant C0C0 and a nontrivial *n*-subpartition {Ti}[n]of VF such that

$$d_F(T_p, T_q) \geq \frac{2}{C_0 n^{5/2}}, \text{ for all } p, q \in [n], \ p \neq q,$$

$$\mu_F(T_p) \geq \frac{1}{2n} \mu_F(\mathcal{V}_F), \text{ for all } p \in [n].$$

The difficulty for the construction of the above *n*-subpartition is to achieve the property (*ii*). That is, we have to find a subpartition which possesses large enough measure. When dF(x,y) is given by the spherical distance $\left\| \frac{F(x)}{\|F(x)\|} - \frac{F(y)}{\|F(y)\|} \right\|$, Theorem 5.6 was proved in [28, 29, Lemma 3.5]. In our situation, we have to deal with the metrics, given in (5.17), of lens spaces or complex projective spaces. We refer the reader to [18] for another interesting decomposition result.

An important ingredient of the proof is the following lemma derived from the random partition theory [20, 30]. Note that a partition of a set *A* can also be considered as a map P:A→2A, where x∈A is mapped to the unique set *P*(*x*) of the partition that contains *x*. A random partition P of *A* is a probability measure νν on a set of partitions of *A*. Then P(x) is understood as a random variable from the probability space to subsets of *A* containing *x*.

LEMMA 5.7

Let A be a subset of the metric space $\mathbb{S}^{2n-1}/\Gamma$ (for *d* recall 5.5). Then for every r>0r>0 and δ∈(0,1)δ∈(0,1), there exists a random partition PP of *A*, i.e., a distribution νν over partitions of *A* such that

1 diam(S)≤r for any *S* in every partition *P* in the support of νν,

2 $\mathbb{P}_\nu\left[B_{r/\alpha}(x) \subseteq \mathcal{P}(x)\right] \geq 1-\delta$ *for all* $x \in A$, *where* $\alpha = 32\log_2(\rho_\Gamma)/\delta$.

We refer to [20, Theorem 3.2] and [30, Lemma 3.11] for the proof, see also [32, Theorem 2.4]. For convenience, we describe briefly the construction of the random partition claimed in Lemma 5.7. Let {xi}[m] be a *r* / 4-net of $\mathbb{S}^{2n-1}/\Gamma$, that is, d(xi,xj)≥r/4, for any i≠j, and $\mathbb{S}^{2n-1}/\Gamma = \bigcup_{i\in[m]} B_{r/4}(x_i)$. Since ($\mathbb{S}^{2n-1}/\Gamma$,d) is compact, *m* is a finite number. For $R \in [r/4, r/2]$ we construct a partition of $(\mathbb{S}^{2n-1}/\Gamma, d)$ as follows. A permutation σσ of the set [*m*] provides an order for all points in the net which is used to define, for every $i \in [m]$,

$$S_i^{R,\sigma} := \left\{x \in \mathbb{S}^{2n-1}/\Gamma \mid x \in B_R(x_i) \text{ and } \sigma(i) < \sigma(j) \text{ for all } j \in [m] \text{ with } x \in B_R(x_j)\right\}.$$

That is, we have $x \in S_i^{R,\sigma}$ if $\sigma(i)$ is the smallest number for which *x* is contained in BR(xi). Then $P^{R,\sigma} = \{S_i^{R,\sigma}\}_{[m]}$ constitutes a partition of $\mathbb{S}^{2n-1}/\Gamma$. Now let σ be a uniformly random permutation of [*m*], and *R* be chosen uniformly random from the interval [*r* / 4, *r* / 2]. These choices define a random partition PP. If we choose *R* uniformly from a fine enough discretization of the interval [*r* / 4, *r* / 2], we can make PP to be finitely supported. In fact, this random partition fulfills the two properties in Lemma 5.7.

Remark 5.8

Lemma 5.7 holds true for any metric space. In particular, the finiteness of the *r* / 4-net is not necessary. This is shown in [30, Lemma 3.11].

Lemma 5.7 leads to the following result. Note that, the property (*ii*) in Lemma 5.7 ensures the existence of at least one subpartition which captures a large fraction of the whole measure.

LEMMA 5.9

On (VF,dF,μF), for any r>0 and δ∈(0,1), there exists a nontrivial subpartition $\{\widehat{S}_i\}_{[m]}$ such that

(i) $\mathrm{diam}(\widehat{S}_i, d_F) \leq r$ *for any* $i \in [m]$,
(ii) $d_F(\widehat{S}_i, \widehat{S}_j) \geq 2r/\alpha$, *where* $\alpha = 32\log_2(\rho_\Gamma)/\delta$,
(iii) $\sum_{l\in[m]} \mu_F(\widehat{S}_i) \geq (1-\delta)\mu_F(\mathcal{V}_F)$.

Proof

Let PP be the random partition on VFVF induced from the one constructed in Lemma 5.7 via the map *F*. Let $I_{B_{r/\alpha}(x)\subseteq\mathcal{P}(x)}$ be the indicator function for the event that $B_{r/\alpha}(x) \subseteq \mathcal{P}(x)$ happens. Then we obtain from Lemma 5.7 (*ii*)

$$\mathbb{E}_{\mathcal{P}}\left(\int_{\mathcal{V}} I_{B_{r/\alpha}(x)\subseteq\mathcal{P}(x)} d\mu_F(x)\right) \geq (1-\delta)\mu_F(\mathcal{V}) \qquad (5.18)$$

by interchanging the expectation and the integral. On the other hand, we have

$$\mathbb{E}_{\mathcal{P}}\left(\int_{\mathcal{V}} I_{B_{r/\alpha}(x)\subseteq \mathcal{P}(x)} d\mu_F(x)\right)$$
$$= \sum_{P\in\mathcal{P}}\sum_{S\in P}\int_S I_{B_{r/\alpha}(x)\subseteq \mathcal{P}(x)} d\mu_F(x)\mathbb{P}_\nu(P) \tag{5.19}$$
$$= \sum_{P\in\mathcal{P}}\sum_{S\in P}\int_{\widehat{S}} d\mu_F(x)\mathbb{P}_\nu(P),$$

where $\widehat{S} := \{x \in S : B_{r/\alpha}(x) \subseteq S\}$.Hence, there exists a partition $P = \{S_i\}_{[m]}$ of $\mathcal{V}_F$ for some natural number *m* such that

$$\sum_{i\in[m]} \mu_F(\widehat{S}_i) \geq (1-\delta)\mu_F(\mathcal{V}). \tag{5.20}$$

This completes the proof.

In order to prove Theorem 5.6, we also need the following result.

LEMMA 5.10

If a subset S⊆V satisfies diam(S∩VF,dF)≤r for some r∈(0,1), then

$$\mu_F(S) \leq \frac{1}{n(1-r^2)}\mu_F(\mathcal{V}). \tag{5.21}$$

Proof

W.l.o.g., we can assume that S⊆V. Using the fact that f1,...,fn are orthonormal, we obtain the following two properties. First, we have

$$\mu_F(\mathcal{V}) = \int_{\mathcal{V}} \sum_{p\in[n]} |f_p|^2 d\mu = n. \tag{5.22}$$

Second, we have for any $\mathbf{z} := (z_1, z_2, \ldots, z_n) \in \mathbb{C}^n$ with $\|\mathbf{z}\| = 1$,,

$$\int_{\mathcal{V}} |\langle \mathbf{z}, F(x)\rangle|^2 d\mu(x) = \int_{\mathcal{V}} \sum_{p,q\in[n]} z_p \overline{z_q} \overline{f_p(x)} f_q(x) d\mu(x) = 1. \tag{5.23}$$

Combining (5.22) and (5.23), we conclude for any y∈Sy∈S,

$$\begin{aligned} \frac{\mu_F(\mathcal{V})}{n} &= \int_{\mathcal{V}} \left| \left\langle \frac{F(y)}{\|F(y)\|}, F(x) \right\rangle \right|^2 d\mu(x) \\ &= \int_{\mathcal{V}} \left| \left\langle \frac{F(y)}{\|F(y)\|}, \frac{F(x)}{\|F(x)\|} \right\rangle \right|^2 d\mu_F(x). \end{aligned} \tag{5.24}$$

Since $|z|^2 \geq (z + \overline{z})^2/4$ for each z∈C, we obtain that for any γ∈Γ:

$$\begin{aligned} \left| \left\langle \frac{F(y)}{\|F(y)\|}, \frac{F(x)}{\|F(x)\|} \right\rangle \right|^2 &= \left| \left\langle \frac{F(y)}{\|F(y)\|}, \gamma \frac{F(x)}{\|F(x)\|} \right\rangle \right|^2 \\ &\geq \frac{1}{4} \left(2 - \left\| \frac{F(y)}{\|F(y)\|} - \gamma \frac{F(x)}{\|F(x)\|} \right\|^2 \right)^2. \end{aligned} \tag{5.25}$$

Recalling (5.17), the definition of dFdF, we arrive at

$$\frac{\mu_F(\mathcal{V})}{n} \geq \int_S \left(1 - \frac{1}{2} d_F(y, x)^2 \right)^2 d\mu_F(x) \geq (1 - r^2)\mu_F(S). \tag{5.26}$$

Proof of Theorem 5.6

With Lemma 5.9 and 5.10 at hand, Theorem 5.6 can be proved similarly as [28, 29, Lemma 3.5], see also [32, Lemma 6.2]. For convenience, we recall it here. Let $\{\widehat{S}_i\}_{[m]}$be the subpartition constructed in Lemma5.9. Then by Lemma 5.10, we have for each $i \in [m]$,

$$\mu_F(\widehat{S}_i) \leq \frac{1}{n(1-r^2)}\mu_F(\mathcal{V}). \tag{5.27}$$

We apply the following procedure to $\{\widehat{S}_i\}_{[m]}$If we can find two sets of the subpartition, say S^i and S^j, such that

$$\mu_F(\widehat{S}_i) \leq \frac{1}{2n}\mu_F(\mathcal{V}), \quad \mu_F(\widehat{S}_j) \leq \frac{1}{2n}\mu_F(\mathcal{V}).$$

then replace them by $\widehat{S}_i \cup \widehat{S}_j$. Thus, when we stop, we obtain the sets $T_1, T_2, \ldots, T_l$ for some number *l*, such that

$$\mu_F(T_i) \leq \frac{1}{n(1-r^2)}\mu_F(\mathcal{V}), \quad \forall i \in [l],$$

and

$$\mu_F(T_i) \geq \frac{1}{2n}\mu_F(\mathcal{V}), \quad \forall i \in [l-1].$$

Setting $r = \frac{1}{3\sqrt{n}}$ and δ=14n, we check that

$$(n-1) \cdot \frac{1}{n(1-r^2)} < 1 - \delta - \frac{1}{2n}. \tag{5.28}$$

This implies that l≥n. Moreover, if we redefine $T_n := \bigcup_{j=n}^{l} T_j$, we have

$$\mu_F(T_n) \geq \frac{1}{2n}\mu_F(\mathcal{V}). \tag{5.29}$$

Thus the subpartition $\{T_i\}_{[n]}$ satisfies the property (*ii*). One can then verify the property (*i*) by Proposition5.2 and Lemma 5.9.

5.4 **Proof of Theorem** 5.1

We first prove the upper bound of (5.1). Let $\{T_i\}_{[n]}$ be the subpartition of VFVF obtained from Theorem 5.6. Choosing $\epsilon = \frac{1}{C_0 n^{5/2}}$, we define the cut-off functions ηpηp as in (5.12) (replacing the set *S* there by TpTp). Then the maps $\Psi_p := \eta_p F,\ p \in [n]$, have pairwise disjoint support. Recalling that $\Psi_p|_{T_p} = F|_{T_p}$, and applying Lemma 5.5 as well as fact (*ii*) of Theorem 5.6, we obtain that for any p∈[n],

$$\begin{aligned} \mathcal{R}^s_\mu(\Psi_p) &\le \left(1+\frac{1}{\epsilon}\right)^2 \frac{\sum_{\{u,v\}\in E} w_{uv}\|F(u) - s_{uv}F(v)\|^2}{\sum_{u\in T_p} \mu(u)\|F(u)\|^2} \\ &\le 2n(1+C_0 n^{5/2})^2 \mathcal{R}^s_\mu(F) \le Cn^6 \mathcal{R}^s_\mu(F), \end{aligned} \tag{5.30}$$

where *C* is an absolute constant. For every p∈[n]p∈[n], the map ΨpΨp has at least one coordinate function ψpψpthat satisfies $\mathcal{R}^s_\mu(\psi_p) \le \mathcal{R}^s_\mu(\Psi_p)$. In particular, we find functions $\psi_p,\ p \in [n]$,], with pairwise disjoint support and an absolute constant *C* such that

$$\mathcal{R}^s_\mu(\psi_p) \le Cn^6 \mathcal{R}^s_\mu(F). \tag{5.31}$$

Now inequality (5.3) and Lemma 4.4 for $\Gamma = S^1_k$ or Lemma 4.8 for Γ=U(1)Γ=U(1) yield the desired upper bound of (5.1).

Now we prove the lower bound of (5.1). Suppose that the *n*-way Cheeger constant hsn(μ)hns(μ) is achieved by the nontrivial *n*-subpartition $\{\widetilde{V}_p\}_{[n]}$ and that the function τ˜p:V˜p→Γ achieves the frustration index ιs(V˜p))for each p∈[n]. Moreover, consider functions f˜p:V→C with pairwise disjoint support given for p∈[n] by:

$$\widetilde{f}_p(u) := \begin{cases} \widetilde{\tau}_p(u), & \text{if } u \in \widetilde{V}_p; \\ 0, & \text{otherwise.} \end{cases} \tag{5.32}$$

By the min–max principle (2.8), we know

$$\lambda_n(\Delta^s_\mu) \leq \max_{a_1,\ldots,a_n} \mathcal{R}^s_\mu(\widetilde{f}_a), \tag{5.33}$$

where the maximum is taken over all complex numbers a1,...,an ∈C such that the linear combination $\widetilde{f}_a := \sum_{p\in[n]} a_p \widetilde{f}_p$ of $\widetilde{f}_1,\ldots,\widetilde{f}_n$ is nontrivial. This implies

$$\sum_{u\in V} \mu(u)|\widetilde{f}_a(u)|^2 = \sum_{p\in[n]} |a_p|^2 \mathrm{vol}_\mu(\widetilde{V}_p). \tag{5.34}$$

We now want to relate (5.33) and (5.34) to the frustration index and the boundary measure. To that direction, we set $B_{uv} := w_{uv}|\widetilde{f}_a(u) - s_{uv}\widetilde{f}_a(v)|^2$ and obtain

$$\sum_{\{u,v\}\in E} B_{uv} = \frac{1}{2}\sum_{p,q\in[n]} \sum_{\substack{u\in\widetilde{V}_p \\ v\in\widetilde{V}_q}} B_{uv} + \sum_{p\in[n]} \sum_{\substack{u\in\widetilde{V}_p \\ v\in V^*}} B_{uv} + \frac{1}{2}\sum_{u,v\in V^*} B_{uv}$$

where $V^* = \left(\bigcup_{p\in[n]} \widetilde{V}_p\right)^c$. For $u, v \in \widetilde{V}_p$, $p \in [n]$, we have

$$|\widetilde{f}_a(u) - s_{uv}\widetilde{f}_a(v)|^2 = |a_p|^2 \cdot |\widetilde{\tau}_p(u) - s_{uv}\widetilde{\tau}_p(v)|^2, \tag{5.35}$$

while for $u \in \widetilde{V}_p$ and $v \in \widetilde{V}_q$ with $p, q \in [n]$ and $p \neq q$ we have

$$|\widetilde{f}_a(u) - s_{uv}\widetilde{f}_a(v)|^2 = |a_p\widetilde{\tau}_p(u) - s_{uv}a_q\widetilde{\tau}_q(v)|^2 \leq 2(|a_p|^2 + |a_q|^2). \tag{5.36}$$

Now the definitions of the frustration index and of the boundary measure yield

$$\begin{aligned} \sum_{\{u,v\}\in E} B_{uv} &\leq \sum_{p\in[n]} |a_p|^2 \left(2\iota^s(\widetilde{V}_p) + 2\left|E(\widetilde{V}_p, \bigcup_{q\neq p} \widetilde{V}_q)\right| + \left|E(\widetilde{V}_p, V^*)\right|\right) \\ &\leq 2\sum_{p\in[n]} |a_p|^2 \left(\iota^s(\widetilde{V}_p) + \left|E(\widetilde{V}_p, \widetilde{V}_p^c)\right|\right). \end{aligned} \tag{5.37}$$

If we now combine the estimates (5.33), (5.34), and (5.37), we arrive at

$$\lambda_n(\Delta_\mu^s) \leq 2\max_{p\in[n]} \phi_\mu^s(\widetilde{V}_p) = 2h_n^s(\mu). \tag{5.38}$$

6 APPLICATION: SPECTRAL CLUSTERING ON ORIENTED GRAPHS AND MIXED GRAPHS

In this section, we discuss an application of the Cheeger inequalities (and their proofs) in the case Γ=S1. These results indicate algorithms to find interesting substructures in an oriented graph or a mixed graph.

6.1 Generalization of Harary's balance theorem

Let us first discuss an equivalent definition of the Cheeger constant hs1(μ)h1s(μ) if Γ=S1kΓ=Sk1. For a nonempty subset $\widetilde{V}$ of V, let $\widetilde{V}_0, \ldots, \widetilde{V}_{k-1}$ be an *ordered k-partition* of $\widetilde{V}$, that is, V˜i are pairwise disjoint sets and their union is V˜V~. In contrast to a nontrivial *k*-partition, all but one V˜i may be empty. We write $\mathscr{V}_k(\widetilde{V})$for an ordered *k*-partition $\widetilde{V}_0, \ldots, \widetilde{V}_{k-1}$ of $\widetilde{V}$.

Given an ordered *k*-partition $\mathscr{V}_k(\widetilde{V})$ of $\widetilde{V} \subseteq V$,we define, for $0 \leq i, j \leq k-1$ and $l \in \mathbb{Z}$,,

$$|E^l(\widetilde{V}_i, \widetilde{V}_j)| := \sum_{u \in \widetilde{V}_i} \sum_{\substack{v \in \widetilde{V}_j \text{ s.t.} \\ s_{uv} = \xi^l}} w_{uv} \tag{6.1}$$

as the (weighted) cardinality of oriented edges with signature ξlξl that begin in V˜i and terminate in V˜j.

DEFINITION 6.1

Let *G* be a finite graph with signature s: $E^{or} \to S^1_k$. For any nonempty subset V˜of *V*, the *k-partiteness ratio* of an ordered *k*-partition $\mathscr{V}_k(\widetilde{V})$ of $\widetilde{V}$·is defined as

$$\beta^s_\mu\left(\mathscr{V}_k(\widetilde{V})\right) = \frac{\frac{1}{2}\sum_{i,j=0}^{k-1}\sum_{l=1}^{k-1}|1-\xi^l|\cdot|E^{i-j+l}(\widetilde{V}_i, \widetilde{V}_j)| + |E(\widetilde{V}, \widetilde{V}^c)|}{\mathrm{vol}_\mu(\widetilde{V})}. \tag{6.2}$$

The *minimal k-partiteness ratio* $\beta^s_\mu(\widetilde{V}, k)$ of $\widetilde{V}$ is defined as

$$\beta_{\mu}^{s}(\widetilde{V},k) := \min_{\mathcal{V}_k(\widetilde{V})} \beta_{\mu}^{s}\left(\mathcal{V}_k(\widetilde{V})\right), \tag{6.3}$$

where the minimum is taken over all ordered k-partitions $\mathcal{V}_k(\widetilde{V})$ of $\widetilde{V}$.

The next goal is to prove that the Cheeger constant for Γ=S1k can also be expressed in terms of the k-partiteness ratio, see Corollary 6.3 below.

LEMMA 6.2

Let G be a finite graph with signature s: $E^{or} \to S_k^1$. For any nonempty $\widetilde{V} \subseteq V$, we have

$$\phi_{\mu}^{s}(\widetilde{V}) = \beta_{\mu}^{s}(\widetilde{V},k). \tag{6.4}$$

Proof

For any function $\tau : V_1 \to S_k^1$, we have a natural k-partition $\mathcal{V}_k(\widetilde{V})$ of $\widetilde{V}$ given by

$$\widetilde{V}_i := \{u \in \widetilde{V} \mid \tau(u) = \xi^i\} \tag{6.5}$$

for i=0,1,...,k−1. We can check that

$$\sum_{\{u,v\}\in\widetilde{E}} w_{uv}|\tau(u) - s_{uv}\tau(v)| = \frac{1}{2}\sum_{i,j=0}^{k-1}\sum_{l=1}^{k-1}|1-\xi^l|\cdot|E^{i-j+l}(\widetilde{V}_i,\widetilde{V}_j)|. \tag{6.6}$$

Observe that the correspondence between the set of S_k^1-valued functions on $\widetilde{V}$ and the set of ordered k-partitions of $\widetilde{V}$ given by (6.5) is one-to-one. Hence, we obtain by definition of the frustration index

$$\iota^s(\widetilde{V}) = \min_{\mathcal{V}_k(\widetilde{V})} \frac{1}{2} \sum_{i,j=0}^{k-1} \sum_{l=1}^{k-1} |1 - \xi^l| \cdot |E^{i-j+l}(\widetilde{V}_i, \widetilde{V}_j)|. \quad (6.7)$$

This proves the lemma.

COROLLARY 6.3

Let G be a finite graph with signature $s : E^{or} \to S_k^1$. Then

$$h_1^s(\mu) = \min_{\emptyset \neq \widetilde{V} \subseteq V} \beta_\mu^s(\widetilde{V}, k). \quad (6.8)$$

This enables us to prove the following structural balance theorem.

THEOREM 6.4

Let G be a finite connected graph with a signature s:Eor→S1ks:Eor→Sk1. Then the following statements are equivalent:

(i)The signature s is balanced.

(ii)There exists an ordered k-partition V0,...,Vk−1, of V such that all edges that begin in Vi and terminate in Vj have signature ξ^{i-j} *for all* $0 \leq i, j \leq k-1$.

Proof

Recall that hs1(μ)=0h1s(μ)=0 if and only if the signature is balanced. The theorem is then a direct consequence of (6.8).

Remark 6.5

Harary's balance theorem [21] states that a signature s:Eor→{±1} is balanced if and only if there exists a bipartition V0,V1, of *V* such that an edge has signature −1 if and only if it has one end point in V0 and one in V1. Theorem 6.4 is a natural generalization of Harary's theorem.

In Figure 1, we schematically illustrate the situation of Theorem 6.4 if k∈{3,4}. The class of edges that begin and terminate in ViVi are represented by one unoriented edge labeled by ξ0=1. For distinct *i, j,* the class of edges with endpoints in ViVi and Vj are represented by an oriented edge that begins in ViVi and terminates in Vj with i<j. These oriented edges are labeled by ξi−j.

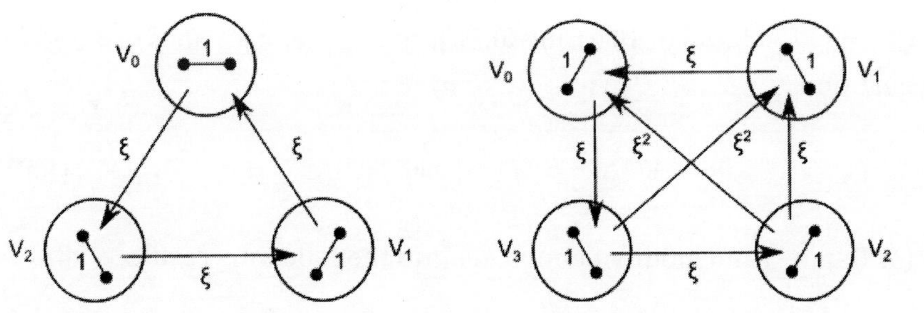

Figure 1. Schematic illustration of Theorem 6.4 for k=3k=3 (*left*) and k=4k=4 (*right*)

6.2 Finding a good substructure

The proof of Cheeger's inequality in Sect. 4, especially Lemma 4.4, actually indicates an algorithm to find a subset $\tilde{V} \subseteq V$ with a constant $\phi^s_\mu(\tilde{V})$ close to the Cheeger constant $h^s_1(\mu)$ of G. In other words, $\phi^s_\mu(\tilde{V})$ is not larger than the upper bound for $h^s_1(\mu)$ given in Cheeger's inequality (Theorem 4.1): for every nonzero function $f: V \to \mathbb{C}$, Lemma 4.4 provides a nonempty subset $\tilde{V} := V^f(\sqrt{t'}) \subseteq V$ satisfying (4.14). If we choose f to be the eigenfunction corresponding to $\lambda_1(\Delta^s_\mu)$ we see that $\tilde{V}$ is a nonempty subset of V with the required property.

Now consider a finite graph G with a k-cyclic signature s. From Lemma 6.2, we know that $\phi^s_\mu(\tilde{V})$ agrees with the minimum of the k-partiteness ratios of all ordered k-partitions $\mathcal{V}_k(\tilde{V})$. Having found a nonempty subset $\tilde{V} := V^f(\sqrt{t'}) \subseteq V$ satisfying (4.14), we explain in this subsection, how to find a finer substructure of $\tilde{V}$, namely an ordered k-partition $\mathcal{V}_k(\tilde{V})$ with a k-partiteness ratio that is at most the upper bound given in (4.14). The precise statement is given in Proposition 6.6 below.

Recall the notation Q^θ_j and $V^f(t)$ of (4.2) and (4.10), respectively. Given $t \in [0,1]$ and $\theta \in [0, 2\pi)$, we define an ordered k-partition $\mathcal{V}_k(V^f(\sqrt{t}, \theta))$ of $V^f(\sqrt{t}, \theta) \subseteq V$ by

$$V^f_j(\sqrt{t}, \theta) := \{u \in V \mid \sqrt{t} \le |f(u)| \text{ and } f(u) \in Q^\theta_j\} \quad (6.9)$$

for $0 \le j \le k-1$ and modify Lemma 4.4 into the following result.

PROPOSITION 6.6

Let s:Eor→S1ks:Eor→Sk1 be a signature of *G*. For any nonzero function f:V→Cf:V→C with $\max_{u\in V}|f(u)|=1$, there exist $t'\in[0,1]\ and\ \theta'\in[0,2\pi)$ such that

$$\beta_{\mu}^{s}\left(\mathcal{V}_k(V^f(\sqrt{t'},\theta'))\right)\le 2\sqrt{2d_\mu \mathcal{R}_\mu^s(f)}. \tag{6.10}$$

Proof

Instead of inequality (4.12), we consider the equality

$$\frac{1}{2\pi}\int_0^{2\pi}\int_0^1 \sum_{\{u,v\}\in E} w_{uv}\left|Y_{\sqrt{t},\theta}(f(u))-s_{uv}Y_{\sqrt{t},\theta}(f(v))\right| dt\,d\theta$$

$$=\frac{1}{2\pi}\int_0^{2\pi}\int_0^1\left(\frac{1}{2}\sum_{i,j=0}^{k-1}\sum_{l=1}^{k-1}\left|1-\xi^l\right|\cdot\left|E^{i-j+l}(W_i,W_j)\right|+\left|E(\widetilde{V},\widetilde{V}^c)\right|\right)dt\,d\theta.$$

where $W_j:=V_j^f(\sqrt{t},\theta)$ and $\widetilde{V}:=V^f(\sqrt{t})$. The remaining proof follows along similar arguments as the ones given in the proof of Lemma 4.4.

This Proposition provides the following spectral clustering algorithm to find an ordered *k*-subpartition of *V*with a *k*-partiteness ratio bounded above by the upper bound in Cheeger's inequality. Firstly, find the eigenfunction f1:V→C corresponding to $\lambda_1(\Delta_\mu^s)$.For convenience, we can normalize f1 such that $\max_{u\in V}|f(u)|=1.$. Secondly, find the required ordered *k*-subpartion from the sets (6.9) by running over fine enough discretizations of the parameters *t* and θ.

6.3 Applications to partially oriented graphs

In this subsection, we consider *mixed graphs* instead of undirected graphs which are studied in scheduling problems, for example [45, 52]. Recall that a *mixed graph* is a graph G=$(V, E_U \cup E_O)$that consists of unoriented edges (the set EUEU) as well as oriented edges (the set EOEO) such that no two vertices u,v∈V form more than one edge of EU∪EOEU∪EO. As mentioned in the introduction, we call such a graph also *partially oriented*. Clearly, a partially oriented graph is an *oriented graph* if and only if EU=∅EU=∅. The algorithm discussed in the previous subsection has interesting applications for partially oriented graphs.

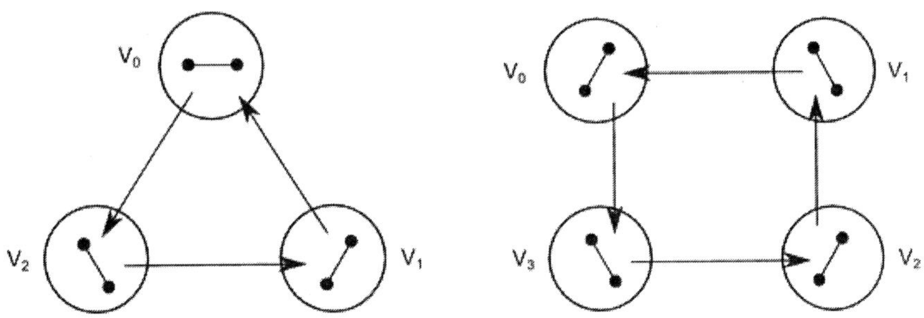

Figure 2. Ideal substructures of partially oriented graphs in case of k=3k=3 (*left*) k=4k=4 (*right*) that are approximated in Sect. 6.3

Given a partially oriented graph G=$(V, E_U \cup E_O)$ and a natural number *k*, we now want to find a nonempty subset V˜⊆V and an ordered *k*-subpartition $\mathcal{V}_k(\tilde{V}) = \{V_0, V_1, \ldots, V_{k-1}\}$ of V˜ which approximates the following ideal substructure:

1 The subset V˜V~ has empty boundary.

2 An $e \in E_U \cup E_O$ with endpoints $u, v \in V_i$ for some $0 \le i \le k-1$ is unoriented, that is, $e \in E_U$.

3 The partially oriented subgraph GV˜induced by V˜has the following *cyclic property*: the only oriented edges of GV˜begin in Vi and end in Vi−1for some 0≤i≤k−1 where we identify V−1 and Vk−1.

Such ideal substructures are schematically illustrated in Figure 2 for k=3 and k=4.

Our approach to this problem is to construct an unoriented graph G=(V,E)G=(V,E) with a *k*-cyclic signature *s*from a given partially oriented graph $G = (V, E_U \cup E_O)$. More precisely, we consider the new edge set $E := E_U \cup E_O$where the orientations in EOEO are dropped and define a signature $s : E^{or} \to S_k^1$: Eor→Sk1 by assigning to every edge {u,v}∈E the value

$$s_{uv} := \begin{cases} 1, & \text{if } \{u, v\} \in E_U; \\ \xi, & \text{if } (u, v) \in E_O; \\ \xi^{-1} & \text{if } (v, u) \in E_O. \end{cases} \quad (6.11)$$

This construction to transform a connected partially oriented graph *G* is set up in such a way that the signature is balanced if and only if *G* has the above ideal structure. Using the eigenfunction of the eigenvalue $\lambda_1(\Delta_\mu^s)$ we apply the spectral clustering algorithm discussed in the Sect. 6.2 to find a *k*-subpartition $\mathcal{V}_k(\widetilde{V})$of some $\widetilde{V} \subseteq V$with *k*-partiteness ratio $\beta_\mu^s(\mathcal{V}_k(V))$at most the upper bound given in Cheeger's inequality. Note that the *k*-partiteness ratio can be viewed as a measure to quantify the quality of an approximation to the ideal case which is achieved if and only if $\beta_\mu^s(\mathcal{V}_k(\widetilde{V})) = 0$. By Corollary 6.3, the *k*-partiteness ratio $\beta_\mu^s(\mathcal{V}_k(\widetilde{V}))$is bounded from below by the Cheeger constant $h_1^s(\mu)$.

We remark that in the special situation were we start with an oriented graph, the ordered *k*-subpartition V0,V1,..., Vk−1V0,V1,...,Vk−1 of *V* approximates an ideal substructure with no edges having both endpoints in ViVi for some 0≤i≤k−1.

These considerations can clearly be extended to obtain multi-way spectral clustering algorithms. Combining the method here with the spectral clustering via metrics on lens spaces in Sect. 5, we can find *n* subgraphs where each subgraph defines a sparse cut and approximates an ideal substructure as described above.

7 MAGNETIC LAPLACIANS ON RIEMANNIAN MANIFOLDS

In this section, we transfer the ideas related to Cheeger constants and Cheeger inequalities from discrete magnetic Laplacians to the Riemannian setting.

Let *M* be a closed connected Riemannian manifold. We consider a real smooth 1-form αααα and the corresponding *magnetic Laplacian* Δ^α on *M*, defined as

$$\Delta^{\alpha} = D^{*} D, \qquad (7.1)$$

where the operator D:=d+iαα, *d* is the exterior differential, maps smooth complex valued functions to smooth complex valued 1-forms and D∗ is the formal adjoint of *D* w.r.t. the L2L2-inner product of functions and 1-forms:

$$\int_M \langle Df, \eta \rangle dx = \int_M f \overline{D^{*}\eta} dx. \qquad (7.2)$$

The 1-form αααα is called the *magnetic potential.* One can check that for any smooth function f:M→C,

$$\Delta^{\alpha} f := \Delta f - 2i \langle df, \alpha \rangle + \left(i d^{*} \alpha + |\alpha|^{2} \right) f, \tag{7.3}$$

where d∗ , Δ:=d∗d is the Laplace–Beltrami operator, ⟨·,·⟩ the Hermitian inner product in the cotangent bundle T∗M induced by the Riemannian metric, and |αα|2:=⟨αα,αα⟩ .

We recall some basic spectral properties of the magnetic Laplacian from [47] (see also [43, Section 4]). The operator Δ^{α} is essentially self-adjoint as an operator defined on smooth complex valued functions (with compact support). Its self-adjoint extension is defined on a dense subset of the Hilbert space L^2(M,C) of complex valued square integrable functions w.r.t the Riemannian measure. In the sequel, we will use the same notation for both, the essentially self-adjoint operator and its closed self-adjoint extension. Since *M* is compact, Δ^{α} has only discrete spectrum, and the eigenvalues can be listed with multiplicity as follows (see [47, Theorem 2.1])

$$0 \leq \lambda_1(\Delta^{\alpha}) \leq \lambda_2(\Delta^{\alpha}) \leq \cdots \nearrow \infty. \tag{7.4}$$

Due to (7.1), the corresponding Rayleigh quotient of a smooth function f:M→C is given by

$$\mathcal{R}^{\alpha}(f) := \frac{\int_M |(d + i\alpha) f|^2 dx}{\int_M |f|^2 dx}. \tag{7.5}$$

The min–max principle (2.8) still holds in this setting. In particular, we have

$$\lambda_1(\Delta^{\alpha}) = \inf_{\substack{f \in C^{\infty}(M,\mathbb{C}) \\ \text{s.t. } f \not\equiv 0}} \mathcal{R}^{\alpha}(f), \tag{7.6}$$

where C^{∞}(M,C) is the set of smooth complex valued functions.

Consider $U(1)$ as a subset $\{z \in C \mid |z|=1\}$ of C and denote the set of smooth maps from M to $U(1)$ by $C^\infty(M,U(1))$. For $\tau \in C^\infty(M,U(1))$ we then define by

$$\alpha_\tau := \frac{d\tau}{i\tau} \tag{7.7}$$

a smooth 1-form. The set $\mathfrak{B} := \{\alpha_\tau \mid \tau \in C^\infty(M, U(1))\}$ has the following characterization due to Shigekawa, [47, Proposition 3.1 and Theorem 4.2]. Since αααт=ααтaαααт=ααтa for a∈R and ααт+ααт′=αатт′, BBis in fact a real vector space.

THEOREM 7.1

(Shigekawa) The following statements are equivalent:

1 $\lambda_1(\Delta^\alpha) = 0;$;

2 $\alpha \in \mathfrak{B};$

3 $d\alpha = 0$ *and* $\int_C \alpha = 0 \mod 2\pi$. for any closed curve *C* in *M*.

This result can be compared with Corollary 4.10: the set BB is comparable to the set of balanced signatures in the discrete setting. Locally, we can find a smooth real-valued function θθ such that $\tau = e^{i\theta}$ and $\alpha_\tau = d\theta$.

In the discrete setting, Laplacians Δ^s_μ with switching equivalent signatures are unitarily equivalent by (2.11) while magnetic Laplacians ΔααΔαα are unitarily equivalent under *gauge transformations* in the smooth setting. Recall that a gauge transformation

$$\alpha \mapsto \alpha + \alpha_\tau \tag{7.8}$$

is associated to any $\tau \in C^{\infty}(M, U(1))$. We have ([47, Proposition 3.2])

$$\overline{\tau}\Delta^{\alpha}\tau = \Delta^{\alpha+\alpha_{\tau}}. \tag{7.9}$$

In particular, if αα∈B, then ΔααΔαα is unitarily equivalent to Δ. In other words, B is the set of magnetic potentials which "can be gauged away".

DEFINITION 7.2

Let αααα be a magnetic potential on *M*. For any nonempty Borel subset Ω⊆M, the *frustration index* ιαα(Ω) of Ω is defined as

$$\iota^{\alpha}(\Omega) = \inf_{\tau \in C^{\infty}(\Omega, U(1))} \int_{\Omega} |(d + i\alpha)\tau| dx = \inf_{\eta \in \mathfrak{B}_{\Omega}} \int_{\Omega} |\eta + \alpha| dx,$$

where $\mathfrak{B}_{\Omega} := \{\alpha_{\tau} | \tau \in C^{\infty}(\Omega, U(1))\}$.

(7.10)

Clearly, the frustration index ι^α (Ω)ιαα(Ω) is invariant under gauge transformations of the potential ααα. Roughly speaking, the frustration index measures how far the potential αααα is from the set BΩ.

For any Borel subset Ω⊆M, we denote by vol(Ω) its Riemannian volume. Its boundary measure area(∂Ω) is defined as

$$\text{area}(\partial\Omega) := \liminf_{r \to 0} \frac{\text{vol}(\Omega_r) - \text{vol}(\Omega)}{r}, \tag{7.11}$$

where Ωr is the open *r*-neighborhood of Ω. Let us denote

$$\phi^{\alpha}(\Omega) := \frac{\iota^{\alpha}(\Omega) + \text{area}(\partial\Omega)}{\text{vol}(\Omega)}. \tag{7.12}$$

DEFINITION 7.3

Let *M* be a closed Riemannian manifold with a magnetic potential αα. The *n-way Cheeger constant* h_n^α is defined as

$$h_n^\alpha := \inf_{\{\Omega_p\}_{[n]}} \max_{p \in [n]} \phi^\alpha(\Omega_p), \tag{7.13}$$

where the infimum is taken over all *n*-subpartitions {Ωp}[n] of *M* with vol(Ωp)>0 for every p∈[n].

In particular, the Cheeger constant hαα1h1αα vanishes if and only if αα∈Bαα∈B. We prove the following lower bound for the first eigenvalue λ1(Δαα)λ1(Δαα).

THEOREM 7.4

Let αααα be a magnetic potential on a closed connected Riemannian manifold *M*. Then we have

$$h_1^\alpha \leq 2\sqrt{2\lambda_1(\Delta^\alpha)}. \tag{7.14}$$

We first prove the following Lemma which is an analogue of Lemma 4.3.

LEMMA 7.5

(Coarea inequality) Let αααα be a magnetic potential on *M*. For any nonzero smooth function f:M→C, we have

$$\int_0^\infty \left(\iota^\alpha(\Omega^f(\sqrt{t})) + \text{area}(\partial\Omega^f(\sqrt{t}))\right) dt \leq 2\sqrt{2} \int_M |f| \cdot |(d + i\alpha)f| dx, \tag{7.15}$$

where we use the notation $\Omega^f(\sqrt{t}) := \{x \in M \mid \sqrt{t} \leq |f(x)|\}$.

Proof

For convenience, we denote f0:=|f| W.l.o.g., we assume that f0(x)>0, for any x∈M. Otherwise, we first consider integration over Ωf(ε) in the right hand side of (7.15), ε>0, and then let ε→0.

For the function *f*, we have the following associated 1-form in BB:

$$\eta_f := \alpha_{\frac{f}{f_0}}. \tag{7.16}$$

Locally, there is a smooth real-valued function θθ such that f/f0=e^iθ^ and ηηf=dθ. Therefore, we have locally

$$|(d + i\alpha)f| = |(d + i\alpha)(f_0 e^{i\theta})| = |df_0 + i f_0(d\theta + \alpha)|. \tag{7.17}$$

This implies that

$$|(d + i\alpha)f| = |df_0 + i f_0(\eta_f + \alpha)|. \tag{7.18}$$

Note that both df0 and f0(ηηf+αα) are real-valued 1-forms. We estimate

$$|(d + i\alpha)f| = \sqrt{|df_0|^2 + |f_0(\eta_f + \alpha)|^2} \geq \frac{1}{\sqrt{2}}\left(|df_0| + |f_0(\eta_f + \alpha)|\right). \tag{7.19}$$

By the coarea formula, we have

$$\int_M f_0 |df_0| \, dx = \int_0^\infty t \cdot \mathrm{area}(\partial\Omega^{f_0}(t)) \, dt. \tag{7.20}$$

We also have

$$\begin{aligned}\int_M f_0^2 \,|\eta_f + \alpha|\, dx &= 2\int_0^{\infty} t \int_{\Omega^{f_0}(t)} |\eta_f + \alpha|\, dx\, dt \\ &\geq \int_0^{\infty} t \int_{\Omega^{f_0}(t)} |\eta_f + \alpha|\, dx\, dt.\end{aligned} \tag{7.21}$$

Combining (7.19), (7.20), and (7.21), we obtain

$$\begin{aligned}\int_M |f| \cdot |(d + i\alpha) f| dx &\geq \frac{1}{2\sqrt{2}} \int_0^{\infty} 2t \left(\text{area}(\partial\Omega^f(t)) + \int_{\Omega^f(t)} |\eta_f + \alpha| dx\right) dt \\ &= \frac{1}{2\sqrt{2}} \int_0^{\infty} \left(\text{area}(\partial\Omega^f(\sqrt{t})) + \int_{\Omega^f(\sqrt{t})} |\eta_f + \alpha| dx\right) dt\end{aligned}$$

Recalling the definition of the frustration index (7.10), this proves the lemma.

Similarly as in Sect. 4 for the discrete setting, we derive the following lemma from the coarea inequality, which is the continuous analogue of Lemma 4.8.

LEMMA 7.6

Let αααα be a magnetic potential on *M*. For any nonzero smooth function f:M→C, there exists $t' \in [0, \max_{x\in M} |f(x)|^2]$ such that

$$\phi^{\alpha}(\Omega^f(\sqrt{t'})) \leq 2\sqrt{2\mathcal{R}^{\alpha}(f)}. \tag{7.22}$$

Proof

First observe that there exists t′t′ such that

$$\phi^{\alpha}(\Omega^f(\sqrt{t'})) \leq \frac{\int_0^{\infty} \left(\iota^{\alpha}(\Omega^f(\sqrt{t})) + \text{area}(\partial\Omega^f(\sqrt{t}))\right) dt}{\int_0^{\infty} \text{vol}(\Omega^f(\sqrt{t})) dt}. \tag{7.23}$$

Note that $\int_M |f(x)|^2 dx = \int_0^\infty \mathrm{vol}(\Omega^f(\sqrt{t}))dt$. Then the lemma follows from applying the coarea inequality and Cauchy–Schwarz inequality.

Theorem 7.4 is proved by applying Lemma 7.6 to the corresponding eigenfunction of λ1(Δ^α) We also have the following higher order Cheeger inequalities for the magnetic Laplacian Δ^α.

THEOREM 7.7

There exists an absolute constant C>0 such that for any closed connected Riemannian manifold *M* with a magnetic potential αααα and n∈Nn∈N, we have

$$h_n^\alpha \leq Cn^3\sqrt{\lambda_n(\Delta^\alpha)}. \tag{7.24}$$

For the proof, first consider Lemma 7.8 below which is an analogue of Lemma 5.5. Let F:M→C be the map given by

$$F(x) = (f_1(x), f_2(x), \ldots, f_n(x)) \in \mathbb{C}^n, \tag{7.25}$$

where fifi are orthonormal eigenfunctions that correspond to the eigenvalues λi(Δ^α) for i∈[n]. The pseudometric dFdF on $M_F := \{x \in M \mid F(x) \neq 0\}$ is defined by (5.11) via

$$d_F(x, y) := \inf_{\gamma \in U(1)} \left\| \frac{F(x)}{\|F(x)\|} - \gamma \frac{F(y)}{\|F(y)\|} \right\|. \tag{7.26}$$

For ϵ>0, the cut-off function ηη from (5.12) is directly transferred to the manifold setting and yields a localized function ηFηF.

LEMMA 7.8

For almost every x∈Mx∈M, we have

$$\|(d+i\alpha)(\eta F)(x)\|^2 \leq 2\left(1+\frac{4}{\epsilon^2}\right)\|(d+i\alpha)F(x)\|^2. \tag{7.27}$$

Proof

If F(x)=0F(x)=0, the estimate (7.27) follows directly from |η|≤1|. We therefore assume F(x)≠0 in the following and set fp,0:=|fp| for every p∈[n]p∈[n]. Then there is a real-valued function θpθp that is defined in a small neighborhood of x∈Mx∈M such that $f_p = f_{p,0}e^{i\theta_p}$. We now obtain at *x*

$$\begin{aligned}
\|(d+i\alpha)(\eta F)\|^2 &= \sum_{p\in[n]} |(d+i\alpha)(\eta f_{p,0}e^{i\theta_p})|^2 \\
&= \sum_{p\in[n]} |f_{p,0}\,d\eta + \eta df_{p,0} + i(\eta f_{p,0})(\alpha + d\theta_p)|^2 \\
&\leq \sum_{p\in[n]} \left(2f_{p,0}^2|d\eta|^2 + 2|\eta|^2|df_{p,0}|^2 + |f_{p,0}(\alpha + d\theta_p)|^2\right) \\
&\leq 2|d\eta|^2 \sum_{p\in[n]} f_{p,0}^2 + 2\sum_{p\in[n]} |df_{p,0} + if_{p,0}(\alpha + d\theta_p)|^2 \\
&= 2|d\eta|^2\|F\|^2 + 2\|(d+i\alpha)F\|^2.
\end{aligned} \tag{7.28}$$

There exist a unit tangent vector $\sigma'(0) \in T_xM$ such that

$$|d\eta(x)| = \lim_{t\to 0} \frac{|\eta(\sigma(t)) - \eta(\sigma(0))|}{t}, \tag{7.29}$$

where σ(t):=expx(tσ′(0)) is the geodesic with σ(0)=x. Since we have

$$|\eta(\sigma(t)) - \eta(\sigma(0))| \leq \frac{1}{\epsilon} \cdot d_F(\sigma(t), \sigma(0)), \tag{7.30}$$

we conclude

$$|d\eta(x)| \cdot \|F(x)\| \leq \frac{1}{\epsilon} \cdot \lim_{t \to 0} \frac{d_F(\sigma(t), \sigma(0)) \cdot \|F(x)\|}{t}. \tag{7.31}$$

Using (7.26) and setting

$$\gamma(t) := e^{i \int_0^t \langle \alpha_{|\sigma(t)}, \sigma'(t) \rangle dt}, \tag{7.32}$$

we obtain

$$\begin{aligned} d_F(\sigma(t), \sigma(0)) \|F(x)\| &\leq \left\| \gamma(t) \frac{F(\sigma(t))}{\|F(\sigma(t))\|} - \frac{F(\sigma(0))}{\|F(\sigma(0))\|} \right\| \cdot \|F(x)\| \\ &= \left\| \frac{G(t)}{\|G(t)\|} - \frac{G(0)}{\|G(0)\|} \right\| \cdot \|G(0)\|, \end{aligned} \tag{7.33}$$

where $G(t) := \gamma(t) F(\sigma(t))$. Now we can carry out similar estimates as in Lemma 5.4. Although we do not know whether ‖G(0)‖ is smaller than ‖G(t)‖, we still obtain

$$\begin{aligned} \left\| \frac{G(t)}{\|G(t)\|} - \frac{G(0)}{\|G(0)\|} \right\| \|G(0)\| &\leq \left\| \frac{\|G(0)\|}{\|G(t)\|} \cdot G(t) - G(t) \right\| + \|G(t) - G(0)\| \\ &\leq 2 \cdot \|G(t) - G(0)\|. \end{aligned} \tag{7.34}$$

Inserting (7.33) and (7.34) into (7.31), we obtain

$$
\begin{aligned}
|d\eta(x)| \cdot \|F(x)\| &\leq \frac{2}{\epsilon} \cdot \lim_{t \to 0} \frac{\|G(t) - G(0)\|}{t} \\
&= \frac{2}{\epsilon} \cdot \lim_{t \to 0} \frac{\sqrt{\sum_{p \in [n]} |\gamma(t) f_p(\sigma(t)) - \gamma(0) f_p(\sigma(0))|^2}}{t} \\
&= \frac{2}{\epsilon} \cdot \sqrt{\sum_{p \in [n]} \left| \lim_{t \to 0} \frac{\gamma(t) f_p(\sigma(t)) - \gamma(0) f_p(\sigma(0))}{t} \right|^2} \\
&= \frac{2}{\epsilon} \cdot \sqrt{\sum_{p \in [n]} |\langle (d + i\boldsymbol{\alpha}) f_p(x), \sigma'(0) \rangle|^2}.
\end{aligned} \tag{7.35}
$$

In the last equality above, we used the fact that $\frac{d\gamma(t)}{dt}|_{t=0} = i\langle \boldsymbol{\alpha}(x), \sigma'(0) \rangle$. Since $|\sigma'(0)| = 1$, we conclude

$$
|d\eta(x)| \cdot \|F(x)\| \leq \frac{2}{\epsilon} \|(d + i\boldsymbol{\alpha}) F(x)\|. \tag{7.36}
$$

Combining (7.36) and (7.28), we finally obtain (7.27).

Note that the pseudometric (7.26) induced from the metric on a complex projective space played an important role in the proof.

Proof of Theorem 7.7

Applying Theorem 5.6 to $(M_F, d_F, \|F(x)\|^2 dx)$, we obtain a subpartition {Ti}[n] of MF, such that

$$
d_F(T_p, T_q) \geq \frac{2}{C_0 n^{5/2}}, \text{ for all } p, q \in [n], p \neq q,
$$

$$
\int_{T_p} \|F(x)\|^2 dx \geq \frac{1}{2n} \int_M \|F(x)\|^2 dx, \text{ for all } p \in [n],
$$

where C0C0 is an absolute constant. Employing further Lemmas 7.6 and 7.8, the proof of the theorem can be done via the same arguments as in Sect. 5.4. □

ACKNOWLEDGMENTS

We like to express our gratitude to Afonso S. Bandeira for pointing out the relation between magnetic and connection Laplacians and useful references. SL is very grateful to Alexander Grigor'yan for inspiring discussions about decompositions of spaces. CL, SL and NP acknowledge the support of the EPSRC Grant EP/K016687/1 "Topology, Geometry and Laplacians of Simplicial Complexes". CL also acknowledges the support of the SFB TRR109 "Discretization in Geometry and Dynamics", the kind hospitality of the Department of Mathematical Sciences of Durham University and of the Grey College.

REFERENCES

1. Alon, N.: Eigenvalues and expanders. Combinatorica **6**(2), 83–96 (1986)

2. Alon, N., Milman, V.: λ1λ1, isoperimetric inequalities for graphs, and superconcentrators. J. Combin. Theory Ser. B **38**(1), 73–88 (1985)

3. Atay, F.M., Liu, S.: Cheeger constants, structural balance, and spectral clustering analysis for signed graphs. arXiv:1411.3530. (2014)

4. Bandeira, A.S., Singer, A., Spielman, D.A.: A Cheeger inequality for the graph connection Laplacian. SIAM J. Matrix Anal. Appl. **34**(4), 1611–1630 (2013)

5. Bauer, F., Jost, J.: Bipartite and neighborhood graphs and the spectrum of the normalized graph Laplacian. Commun. Anal. Geom. **21**(4), 787–845 (2013)

6. Bauer, F., Keller, M., Wojciechowski, R.K.: Cheeger inequalities for unbounded graph Laplacians. J. Eur. Math. Soc. (JEMS) **17**(2), 259–271 (2015)

7. Cheeger, J.: A lower bound for the smallest eigenvalue of the Laplacian, Problems in analysis (Papers dedicated to Salomon Bochner, 1969), pp. 195–199, Princeton Univ. Press, Princeton, NJ (1970)

8. Coifman, R., Weiss, G.: Analyse harmonique non-commutative sur certains espaces homogènes. Lecture Notes in Mathematics, vol. 242. Springer, Berlin-New York (1971)

9. Colin de Verdière, Y., Torki-Hamza, N., Truc, F.: Essential self-adjointness for combinatorial Schrödinger operators III-Magnetic fields. Ann. Fac. Sci. Toulouse Math. (6) 20(3), 599–611 (2011)

10. Desai, M., Rao, V.: A characterization of the smallest eigenvalue of a graph. J. Graph Theory **18**(2), 181–194 (1994)

11. Dodziuk, J.: Difference equations, isoperimetric inequality and transience of certain random walks. Trans. Am. Math. Soc. **284**(2), 787–794 (1984)

12. Dodziuk, J., Mathai, V.: Kato's inequality and asymptotic spectral properties for discrete magnetic Laplacians. The ubiquitous heat kernel (Boulder, CO, 2003), 69–81, Contemp. Math., 398, Am. Math. Soc. Providence, RI (2006)

13. Erdős, L.: Rayleigh-type isoperimetric inequality with a homogeneous magnetic field. Calc. Var. **4**, 283–292 (1996)

14. Frank, R.L., Laptev, A., Molchanov, S.: Eigenvalue estimates for magnetic Schrödinger operators in domains. Proc. Am. Math. Soc. **136**(12), 4245–4255 (2008)

15. Funano, K.: Eigenvalues of Laplacian and multi-way isoperimetric constants on weighted Riemannian manifolds. arXiv:1307.3919v1. (2013)

16. Funano, K., Shioya, T.: Concentration, Ricci curvature, and eigenvalues of Laplacian. Geom. Funct. Anal.**23**(3), 888–936 (2013)

17. Golénia, S.: Hardy inequality and asymptotic eigenvalue distribution for discrete Laplacians. J. Funct. Anal. **266**, 2662–2688 (2014)

18. Grigor'yan, A., Netrusov, Y., Yau, S.-T.: Eigenvalues of elliptic operators and geometric applications. Surveys in differential geometry. Vol. IX, pp. 147–217. Int. Press, Somerville, MA (2004)

19. Gross, J.L.: Voltage graphs. Discr. Math. **9**, 239–246 (1974)

20. Gupta, A., Krauthgamer, R., Lee, J.R.: Bounded geometries, fractals, and low-distortion embeddings In: 2003 IEEE 44th annual symposium on foundations of computer science-FOCS 2003. IEEE Computer Soc., pp. 534–543. Washington, DC(2003)

21. Harary, F.: On the notion of balance of a signed graph. Michigan Math. J. **2**(2), 143–146 (1953)

22. Harary, F.: On the measurement of structural balance. Behav. Sci. **4**, 316–323 (1959)

23. Harary, F., Palmer, E.: Enumeration of mixed graphs. Proc. Am. Math. Soc. **17**, 682–687 (1966)

24. Hinz, M., Teplyaev, A.: Dirac and magnetic Schrödinger operators on fractals. J. Funct. Anal. **265**, 2830–2854 (2013)

25. Hoory, S., Linial, N., Wigderson, A.: Expander graphs and their applications. Bull. Am. Math. Soc. **43**(4), 439–561 (2006)

26. Jost, J.: Riemannian Geometry and Geometric Analysis, 4th edn. Universitext, Springer, Berlin (2005)

27. Kwok, T.-C., Lau, L.-C., Lee, Y.-T., Oveis Gharan, S., Trevisan, L.: Improved Cheeger's inequality: analysis of spectral partitioning algorithms through higher order spectral gap. In: STOC'13-proceedings of the 2013 ACM symposium on theory of computing, 11–20, ACM, New York (2013)

28. Lee, J. R., Oveis Gharan, S., Trevisan, L.: Multi-way spectral partitioning and higher-order Cheeger inequalities. In: STOC'12-proceedings of the 2012 ACM symposium on theory of computing, 1117–1130, ACM, New York (2012)

29. Lee, J. R., Oveis Gharan, S., Trevisan, L.: Multiway spectral partitioning and higher-order Cheeger inequalities. J. ACM **61**(6), 37:1–30 (2014)

30. Lee, J.R., Naor, A.: Extending Lipschitz functions via random metric partitions. Invent. Math. **160**(1), 59–95 (2005)

31. Lieb, E., Loss, M.: Fluxes, Laplacians, and Kasteleyn's theorem. Duke Math. J. **71**(2), 337–363 (1993)

32. Liu, S.: Multi-way dual Cheeger constants and spectral bounds of graphs. Adv. Math. **268**, 306–338 (2015)

33. Liu, S.: An optimal dimension-free upper bound for eigenvalue ratios. arXiv:1405.2213. (2014)

34. Liu, S., Peyerimhoff, N.: Eigenvalue ratios of nonnegatively curved graphs. arXiv:1406.6617. (2014)

35. Liu, S., Peyerimhoff, N., Vdovina, A.: Signatures, lifts and eigenvalues of graphs. arXiv:1412.6841. (2014)

36. von Luxburg, U.: A tutorial on spectral clustering. Stat. Comput. **17**(4), 395–416 (2007)

37. Marcus, A.W., Spielman, D.A., Srivastava, N.: Interlacing families I: bipartite Ramanujan graphs of all degrees. In: 2013 IEEE 54th annual symposium on foundations of computer science - FOCS 2013. IEEE Computer Soc., pp 529–537. Los Alamitos, CA (2013)

38. Marcus, A.W., Spielman, D.A., Srivastava, N.: Interlacing families I: bipartite Ramanujan graphs of all degrees. Ann. Math. **182**, 307–325 (2015)

39. Miclo, L.: On eigenfunctions of Markov processes on trees. Probab. Theory Rel Fields **142**(3–4), 561–594 (2008)

40. Miclo, L.: On hyperboundedness and spectrum of Markov operators. Invent. Math. **200**(1), 311–343 (2015)

41. Morame, A., Truc, F.: Counting function of the embedded eigenvalues for some manifold with cusps, and magnetic Laplacian. Math. Res. Lett. **19**(2), 417–429 (2012)

42. Nicaise, S.: Spectre des réseaux topologiques finis. Bull. Sci. Math. (2) **111**(4), 401–413 (1987)

43. Paternain, G.P.: Schrödinger operators with magnetic fields and minimal action functionals. Israel J. Math. **123**, 1–27 (2001)

44. Post, O.: Spectral analysis of metric graphs and related spaces. In: Arzhantseva, G., Valette, A. (eds.) Limits of graphs in group theory and computer science, pp. 109–140. EPFL Press, Lausanne, (2009)

45. Ries, B.: Coloring some classes of mixed graphs. Discr. Appl. Math. **155**(1), 1–6 (2007)

46. Sadeghi, K., Lauritzen, S.: Markov properties for mixed graphs. Bernoulli **20**(2), 676–696 (2014)

47. Shigekawa, I.: Eigenvalue problems for the Schrödinger operator with the magnetic field on a compact Riemannian manifold. J. Funct. Anal. **75**(1), 92–127 (1987)

48. Shubin, M.A.: Discrete magnetic Laplacian. Commun. Math. Phys. **164**(2), 259–275 (1994)

49. Shubin, M.A.: Essential self-adjointness for semi-bounded magnetic Schrödinger operators on non-compact manifolds. J. Funct. Anal. **186**(1), 92–116 (2001)

50. Simon, B., Høegh-Krohn, R.: Hypercontractive semigroups and two dimensional self-coupled Bose fields. J. Funct. Anal. **9**, 121–180 (1972)

51. Singer, A., Wu, H.-T.: Vector diffusion maps and the connection Laplacian. Commun. Pure Appl. Math.**65**(8), 1067–1144 (2012)

52. Sotskov, Y.N.: Scheduling via mixed graph coloring. In: Operations research proceedings 1999 (Magdeburg), 414–418. Springer, Berlin (2000)

53. Sunada, T.: A discrete analogue of periodic magnetic Schrödinger operators. Geometry of the spectrum (Seattle, WA, 1993), 283–299, Contemp. Math., 173, Am. Math. Soc., Providence, RI (1994)

54. Trevisan, L.: Max cut and the smallest eigenvalue. In: STOC'09-Proceedings of the 2009 ACM international symposium on theory of computing, 263–271, ACM, New York (2009)

55. Trevisan, L.: Max cut and the smallest eigenvalue. SIAM J. Comput. **41**(6), 1769–1786 (2012)

56. Vannimenus, J., Toulouse, G.: Theory of the frustration effect: II. Ising spins on a square lattice. J. Phys. C: Solid State Phys. **10**, L537 (1977)

57. Wang, F.-Y.: Criteria of spectral gap for Markov operators. J. Funct. Anal. **266**, 2137–2152 (2014)

58. Zaslavsky, T.: Signed graphs. Discr. Appl. Math. **4**(1), 47–74 (1982)

59. Zaslavsky, T.: Matrices in the theory of signed simple graphs. Advances in discrete mathematics and applications (Mysore, 2008). Ramanujan Math. Soc. Lect. Notes Ser. 13. RamanujanMath. Soc., pp. 207–229. Mysore (2010)

60. Zhang, X.-D., Li, J.-S.: The Laplacian spectrum of a mixed graph. Linear Algebra Appl. **353**, 11–20 (2002)

CHAPTER 9

Bending of thin periodic plates

Mikhail Cherdantsev[1] · Kirill Cherednichenko[2]

1 School of Mathematics, Cardiff University, Senghennydd Road, Cardiff CF24 4AG, UK

2 Department of Mathematical Sciences, University of Bath, Claverton Down, Bath BA2 7AY, UK

Abstract

We show that nonlinearly elastic plates of thickness $h\to0h\to0$ with an $\varepsilon\varepsilon$-periodic structure such that $\varepsilon-2h\to0\varepsilon-2h\to0$exhibit non-standard behaviour in the asymptotic two-dimensional reduction from three-dimensional elasticity: in general, their effective stored-energy density is "discontinuously anisotropic" in all directions. The proof relies on a new result concerning an additional isometric constraint that deformation fields must satisfy on the microscale.

1 INTRODUCTION

Understanding the behaviour of elastic plates (and, more generally, thin elastic structures) from the rigorous mathematical point of view has attracted much attention of applied analysts over the recent years. The related activity was initiated by the papers [1] and [15], which were followed by the work [8] by Friesecke, James and Müller concerning plate deformations *u* with finite "bending energy". The paper [8] appeared alongside the thesis [21] by Pantz and was followed by a study of other energy scalings [9]. It puts forward the idea that homogeneous plates of thickness *h*, viewed as three-dimensional nonlinearly elastic bodies, afford a special compactness argument for sequences of deformation gradients $\nabla u \nabla u$ with elastic energy of order $O(h3)O(h3)$ as $h \to 0.h \to 0$. This argument is based on a new "rigidity estimate" [8] concerning the distance of local values of $\nabla u \nabla u$ from the group of rotations $SO(3).SO(3)$. The observation that this bounds the distance from a constant rotation field (which depends on *u*) allowed the authors of [8] to show that the "limit" elastic energy functional as $h \to 0h \to 0$ is given by

$$E_{\mathrm{lim}}(u) := \frac{1}{12}\int_{\omega} Q_2(\mathrm{II}(u)), \quad u \in H^2_{\mathrm{iso}}(\omega), \qquad (1)$$

where $\omega \subset R2$ represents the mid-surface of the undeformed plate, $II(u)=(\nabla' u)T\nabla'(\partial 1u \wedge \partial 2u)$, $\nabla'=(\partial 1,\partial 2)$, is the matrix of the second fundamental form of an isometric surface $u:\omega \to R3$, and $Q2$ is a quadratic form derived via dimension reduction from a quadratic form appearing in the process of linearisation of elastic properties of the material in the small-strain regime.

From the point of view of applications to real-world materials, it seems reasonable to ask in what way the above result is affected by a possible inhomogeneity of material properties of the plate in the directions tangential to its mid-surface. One can imagine, for instance, that the plate has a periodic structure of period $\varepsilon>0$ and try to replace the $\varepsilon,h\varepsilon,h$-dependent family of $h-3$ scaled energy functionals by an "effective" functional in the sense of variational convergence [6]. Such analysis should reveal, in particular, whether the behaviour of the plate really depends on the relative orders of smallness of the parameters $\varepsilon\varepsilon$ and *h*. Work on this programme was started in the thesis [17] by Neukamm, where a series of similar questions was addressed for a periodically inhomogeneous rod, i.e. a one-dimensional analogue of a plate. Further, a recent paper [14] has investigated the behaviour of periodic plates in the cases when $h\gg\varepsilon h\gg\varepsilon$ and $h\sim\varepsilon h\sim\varepsilon$. This was followed by the work [23], where the case $\varepsilon 2\ll h\ll\varepsilon$ is addressed. In the present paper we develop an approach (see Sects. 3, 4) that in our view simplifies the derivation of the corresponding $\Gamma\Gamma$-limit, via a "smoothing" approximation procedure that precedes the two-scale compactness argument, see Sect. 3 of the present work. Smoothing is known to be useful in the asymptotic analysis of sequences of solutions to parameter-dependent PDE (see e.g. [10, 25]). In the present work we exploit similar considerations in the asymptotic analysis of bounded-energy sequences, where the energy is represented by an integral functional. Our smoothing approach replaces the approximation result [14, Lemma 3.8], when $h\gg\varepsilon$ or $h\sim\varepsilon$, , and an additional technical statement [23, Lemma 3.7], when $\varepsilon 2\ll h\ll\varepsilon$, by a two-scale compactness theorem for the second gradients of smooth approximations (Theorem 3 below). Remarkably, in the situation where the deformation $uh\in H1(\Omega)$ has no second derivatives, the smoothing procedure takes us to a setting where the second derivatives exist and are bounded in the L2-norm, see Lemma 2.

The added value of our approach, however, is revealed through its ability to deal with the more difficult case $h\ll\varepsilon 2$, , which has remained open until now. An additional key observation in handling this case is that the determinant of the matrix of second derivatives of the smoothed approximation can be rewritten in a form amenable to the use of a compensated compactness argument (see proof of Theorem 5) in order to derive an additional, "fast-scale", isometry constraint on the limit finite-energy deformations, see (2). It also perhaps worth mentioning that the smoothing approach yields a somewhat shorter route to the statement of compactness of finite-energy sequences in the case of homogeneous plates considered in [8].

Notably, a recent paper [18] contains an analysis of "zero-thickness" plates, where functionals of the "limit" form (1) with explicit εε-periodic *x*-dependence of Q2Q2 are studied in the limit ε→0.ε→0. Our results are consistent with those of [18], in the sense of offering an alternative route to the same "supercritically-thin plate" limit functional. However, while the authors of [18] work with a two-dimensional formulation from the outset, our derivation is in the spirit of "dimension reduction" from the full three-dimensional problem. In addition, our approach offers a shorter route to the proofs of [18], which is checked directly.

All of the above earlier works follow a traditional methodology of Γ-convergence relying solely on convergence properties of sequences with bounded energy combined with "structure theorems" for combinations of their gradients, without analysing their asymptotic structure as a two-scale series in powers of the parameters ε,h. Our approach, on the contrary, is guided by an analysis of the original sequence of energy functionals via two-scale power series expansions (For a detailed discussion of the viewpoint provided by the method of asymptotic expansions, see Appendix 2). For "supercritical" scalings h≪ε2 it suggests in particular that, to the leading order, the approximating ("exact" or "recovery") sequences have to satisfy an additional constraint

$$\det(\mathrm{II} + \nabla_y^2 \psi) = 0, \tag{2}$$

where $\psi = \psi(x', y),\ x' \in \omega,\ y \in Y := [0, 1)^2$, is a term that appears in the two-scale limit of the original sequence of deformations and is responsible for the behaviour of the plate on εε-scale. One key aspect that facilitates this approach is that the limit stored-energy function Q2 is quadratic with respect to the second fundamental form of the surface, which makes the analysis of ε,hε,h-dependent plate energies somewhat amenable to a perturbation technique.

The observation that the space of deformations with finite bending energy acquires the constraint (2), combined with a characterisation of isometric embeddings by Pakzad [20] and the earlier ansatz of [8], see (77), allows us to carry out a rigorous proof of variational convergence. Thanks to the two-scale asymptotic structure, the presence of the second gradient of the corrector ψψ in the two-scale limit of deformation sequences provides a natural construction for the recovery sequences, cf. (37) for the moderate regime. The supercritical regime h≪ε2 is studied in Sect. 7, which contains our main result, Theorem 9: the limit energy functional is given by

$$E^{\rm sc}_{\rm hom} = \frac{1}{12}\int_{\omega} Q^{\rm sc}_{\rm hom}\big(\mathrm{II}(u)\big),$$

where

$$Q^{\rm sc}_{\rm hom} = \min \int_{Y} Q_2(y, \mathrm{II} + \nabla_y^2 \psi)dy, \quad \psi \in H^2_{\rm loc}(\mathbb{R}^2)\ Y - \text{periodic, subject to (2).}$$

The isometry constraint (2) implies, in particular, that $Q^{\rm sc}(\mathrm{II})$ is "discontinuously anisotropic in all directions of bending" as a function of the macroscopic deformation gradient ∇u, see Theorem 8 and the subsequent discussion. To the best of our knowledge, this is a new phenomenon for nonlinearly elastic plates.

Finally, we remark that the analysis of the "critical" scaling h~ε^2 is currently open.

Throughout the text Id denotes the d×d identity matrix, d=2,3. Unless indicated otherwise, we denote by *C* a positive constant whose precise value is of no importance and may vary between different formulae. We use the notation ∂i, i=1,2,3 for the partial derivative with respect to xixi, ∂yi, i=1,2 for the partial derivative with respect to yiyi, and ∇y for the gradient with respect to *y*.

2 SETTING OF THE PROBLEM

Let ωω be a bounded convex1 domain in R2R2, and let $\Omega_h := \omega \times (-\frac{h}{2}, \frac{h}{2})$ be a reference configuration of an undeformed elastic plate, where *h* is a small positive parameter, 0<h≪1. A deformation *u* of the plate Ωh is an H1-function from ΩhΩh into R3. We assume that the material properties of the plate Ωh vary over ω with a period ε, which goes to zero simultaneously with *h*. For a plate made of a hyperelastic material (see e.g. [4]), the elastic energy of a deformation *u* is given by the functional

$$\int_{\Omega_h} W(\varepsilon^{-1}x', \nabla u)dx,$$

where x′:=(x1,x2)∈ω. The stored-energy density *W*(*y*, *F*) is assumed to be a Carathéodory function (i.e. measurable in y∈Y:=$[0,1)^2$ and continuous in F∈$R^{3\times3}$), periodic with respect to y∈Y, and to satisfy the standard conditions of nonlinear elasticity (see e.g. [4]):

The density W is frame-indifferent, i.e. W(y,RF)=W(F) for any R∈SO(3) and any F∈$R^{3\times3}$, where SO(3) is a group of rotation matrices in $R^{3\times3}$.

The identity deformation u(x)=x is a natural state, i.e. W(y,I3)=minF W(y,F)=0.

There exists a constant C>0 such that for all y∈Y F∈$R^{3\times3}$ he inequality

$$W(y, F) \geq C\, \mathrm{dist}^2(F, \mathrm{SO}(3)) \tag{3}$$

holds.

Additionally we assume that (cf. [3, 8, 14])

(D) The density W, as a function of the deformation gradient, admits a quadratic expansion at the identity, i.e. there exists a non-negative quadratic form Q_3 on $R^{3\times3}$ such that

$$W(y, \mathcal{I}_3 + G) = Q_3(y, G) + r(y, G) \tag{4}$$

for all G∈$R^{3\times3}$, where

$$r(y, G) = o(|G|^2) \quad \text{as} \quad |G| \to 0 \quad \text{uniformly with respect to } y \in Y. \tag{5}$$

We assume additionally that Q_3 is measurable in y∈Y for all G, |G|=1, so that r is a Carathéodory function.

We next rescale the transverse variable x3x3 in order to work on a fixed domain. Namely, we multiply the transverse variable by h^{-1} so that the new variable belongs to the interval $I := (-\frac{1}{2}, \frac{1}{2})$. We reassign the notation x_3 to the new variable (this should not cause any misunderstanding since we will always deal only with the rescaled domain in the future), so x∈Ω:=ω×I. We also use the rescaled gradient

$$\nabla_h := \left(\nabla', \frac{1}{h}\frac{\partial}{\partial x_3}\right) := \left(\frac{\partial}{\partial x_1}, \frac{\partial}{\partial x_3}, \frac{1}{h}\frac{\partial}{\partial x_3}\right)$$

so that the rescaled functional is given by

$$E_h(u) := \int_\Omega W(\varepsilon^{-1}x', \nabla_h u)dx,$$

where $u \in H^1(\Omega)$. We consider sequences of deformations uhuh in the bending regime, that is such that

$$E_h(u_h) \leq Ch^2. \tag{6}$$

(In the non-rescaled setting this corresponds to the energies of order h^3). In this paper we focus on the case when h≪ε. We consider εε as a function of *h*, i.e. *h* is our "main" parameter and εε is the "dependent" parameter.

The non-degeneracy property (3) and the assumption (6) imply that

$$\int_\Omega \operatorname{dist}^2\big(\nabla_h u_h, SO(3)\big)dx \leq Ch^2. \tag{7}$$

In what follows, we say that a sequence satisfying (7) has finite bending energy.

3 TWO-SCALE COMPACTNESS AND SECOND GRADIENTS

The two cases when εε is of order *h* and h≫εh≫ε were considered in [14]. In these regimes the behaviour of the mid-surface of the plate is essentially macroscopic and the microscale only plays a role away from the mid-surface. In our case the microscopic behaviour of the mid-surface of the plate is very important. This is due to the fact that for each εε-cell the corresponding piece of the plate itself behaves like a miniature plate. This "micro-plate" behaviour becomes even more dominant when h≪ε2h≪ε2. In order to study this property we employ the method of two-scale convergence. The structure of the limit functional in the case of a homogeneous plate, namely, the fact that it is basically defined on the second gradient of the limit surface, suggests that the information about the microscopic behaviour should also appear in the form of a second gradient. The problem is that the original functional is defined only on H1H1-functions, and how one would obtain the second derivatives with respect to the fast variable *y* in the limit is not clear at the outset. However, while this is not possible in general, in the bending regime deformations possess an additional property which in a certain sense is equivalent to the existence of the second gradient, which we explain in detail in what follows.

Let us recall some results from [8]. We denote

$$Y_{a,h} := a + hY, \quad a \in h\mathbb{Z}^2, \quad \omega_h := \bigcup_{Y_{a,h} \subset \omega} Y_{a,h}.$$

THEOREM 1

Suppose that a sequence uhuh in H1(Ω)H1(Ω) has finite bending energy. Then:

- Up to a subsequence

$$\nabla_h u_h \to (\nabla' u, n) \quad \textit{strongly in } L^2(\Omega),$$

where u=u(x′) belongs to the space $H^2_{\text{iso}}(\omega)$ of isometric immersions, i.e. such maps u:ω→R^3 that and n:=u,1∧u,2:

- There exists a piecewise constant map Rh:ωh→SO(3) such that

$$\int_{\omega_h \times I} |\nabla_h u_h - R_h|^2 \, dx \le Ch^2$$

for some C>0independent of *h*;

- The difference between the values of R_h in each pair of neighbouring *h*-cells is small in the sense that

$$\int_{\omega_h} \max_{\zeta \in \Upsilon_{x'}} \left| R_h(x' + \zeta) - R_h(x') \right|^2 dx' \le Ch^2,$$

where C>0 is independent of *h*, and ϒx′ consists of those ζ∈{−h,0,h}^{2} for which $Y_{a+\zeta,h}$⊂ωh for the value a∈hZ^2 such that x′∈Ya,h.

Apart from the strong compactness of the gradients ∇huh, which is true in our setting as well as for homogeneous plates, this theorem implies that oscillations of the gradients on the *h*-scale are bounded. The third property in Theorem 1 implies that the difference between values of RhRh in neighbouring *h*-cells divided by *h* (which basically is the difference quotient) is bounded on average and a similar statement can be formulated for ∇huh due to the second part of the theorem. Hence one can try to mollify uh,uh, expecting that the second gradient of the mollification will be bounded. In what follows we always assume that uhuh is a sequence with finite bending energy.

Let $\varphi \in C_0^\infty(B_1),\ \varphi \ge 0,$ be radially symmetric, where B_1 is a unit disc centred at the origin, ∫B1φ=1, and for each h>0 denote by $\varphi_h(x') := h^{-2}\varphi(h^{-1}x'),\ x' \in \mathbb{R}^2$, the corresponding mollifier. Let ω′ be a domain whose closure is contained in ω. For each sufficiently small h>0, consider the function $\widetilde{u}_h = \widetilde{u}_h(x'),\ x' \in \omega'$ defined as the result of the simultaneous mollification of uhuh with φhφh and averaging with respect to the variable x_3:

$$\widetilde{u}_h(x') := \int_I \int_{\mathbb{R}^2} \varphi_h(\xi - x') u_h(\xi, x_3)\, d\xi dx_3 = \int_I \int_{\mathbb{R}^2} \varphi_h(\xi) u_h(\xi + x', x_3)\, d\xi dx_3, \quad x' \in \omega'. \tag{8}$$

(The above integral is well defined in ω′ω′ for sufficiently small *h*, since in this case the support of φh(ξ−x′) is contained in ω). It is well known that the mollification u˜hu∼h is infinitely smooth. We will need to estimate in H_1 the difference between u˜hu∼h and the average of uhuh with respect to x_3,

$$\overline{u}_h(x') := \int_I u_h(x) dx_3, \quad x' \in \omega',$$

as well as its derivatives up to the second-order. Such estimates are obtained in the next statement.

THEOREM 2

Let ω′ω′ be a domain whose closure is contained in ωω. The mollification (8) satisfies the following inequalities with a constant C>0C>0 independent of ω′ω′ and *h*:

$$\|\widetilde{u}_h - \overline{u}_h\|_{H^1(\omega')} \le Ch, \tag{9}$$

$$\|\nabla'^2\widetilde{u}_h\|_{L^2(\omega')} \leq C,$$
$$\|\nabla'\widetilde{u}_h\|_{L^\infty(\omega')} \leq C. \tag{10}$$

Proof

In the subsequent estimates we use a generalised Minkowski inequality (also sometimes referred to as the Minkowski integral inequality):

$$\left[\int_{\mathcal{Y}}\left|\int_{\mathcal{X}} F(x,y)d\mu_{\mathcal{X}}(x)\right|^p d\mu_{\mathcal{Y}}(y)\right]^{1/p} \leq \int_{\mathcal{X}}\left[\int_{\mathcal{Y}}\left|F(x,y)\right|^p d\mu_{\mathcal{Y}}(y)\right]^{1/p} d\mu_{\mathcal{X}}(x).$$

where X and Y are measure spaces with measures μX(x) and μY(y) respectively, F:X×Y→R is measurable and 1<p<∞ , see e.g. [11]. We obtain the L_2-estimate in (9) as follows:

$$= \left(\int_{\omega'}\left(\int_{\mathbb{R}^2}\varphi_h(\xi)\left(\overline{u}_h(\xi+x')-\overline{u}_h(x')\right)d\xi\right)^2 dx'\right)^{1/2}$$
$$\leq \int_{\mathbb{R}^2}\left(\int_{\omega'}\left(\varphi_h(\xi)\left(\overline{u}_h(\xi+x')-\overline{u}_h(x')\right)\right)^2 dx'\right)^{1/2} d\xi \leq \max_{|\xi|\leq h}\left\|\overline{u}_h(\xi+x')-\overline{u}_h(x')\right\|_{L^2(\omega')}$$
$$= \max_{|\xi|\leq h}\left\|\int_0^1 \nabla'\overline{u}_h(t\xi+x')\cdot\xi\, dt\right\|_{L^2(\omega')} \leq \max_{|\xi|\leq h}|\xi|\int_0^1\left\|\nabla'\overline{u}_h(t\xi+x')\right\|_{L^2(\omega')} dt$$
$$\leq Ch\left\|\nabla'\overline{u}_h(x')\right\|_{L^2(\omega)} \leq Ch.$$

Now we estimate the difference of the gradients of $\widetilde{u}_h$ and $\overline{u}_h$. First notice that by the properties of mollification, $\nabla'\widetilde{u}_h$ equals to the mollification of . Second, one has

$$\|\nabla'\overline{u}_h - \widehat{R}_h\|_{L^2(\omega')} \leq \|\nabla' u_h - \widehat{R}_h\|_{L^2(\Omega')} \leq Ch. \tag{11}$$

where $\widehat{R}_h$stands for the first two columns of the matrix RhRh, and Ω′:=ω′×I. The bound (11) is a simple consequence of application of the generalised Minkowski inequality and the Hölder inequality. Further, the following estimate holds:

$$\begin{aligned}
&\|\nabla'\widetilde{u}_h - \nabla'\overline{u}_h\|_{L^2(\omega')} \\
&= \left\| \int_{\mathbb{R}^2} \varphi_h(\xi)\left(\nabla'\overline{u}_h(\xi + x') - \nabla'\overline{u}_h(x')\right) d\xi \right\|_{L^2(\omega')} \\
&\leq \int_{\mathbb{R}^2} \left\|\varphi_h(\xi)\left(\nabla'\overline{u}_h(\xi + x') - \nabla'\overline{u}_h(x')\right)\right\|_{L^2(\omega')} d\xi \\
&\leq \max_{|\xi|\leq h} \left\|\nabla'\overline{u}_h(\xi + x') - \nabla'\overline{u}_h(x')\right\|_{L^2(\omega')} \\
&\leq \max_{|\xi|\leq h} \left\|\nabla'\overline{u}_h(\xi + x') - \widehat{R}_h(x' + \xi)\right\|_{L^2(\omega')} + \max_{|\xi|\leq h} \left\|\widehat{R}_h(\xi + x') - \widehat{R}_h(x')\right\|_{L^2(\omega')} \\
&\quad + \left\|\widehat{R}_h(x') - \nabla'\overline{u}_h(x')\right\|_{L^2(\omega')} \leq Ch.
\end{aligned} \tag{12}$$

where we use the generalised Minkowski inequality and Theorem 1.

Next we estimate the second gradient of u $\widetilde{u}_h$, given by

$$\nabla'^2\widetilde{u}_h = \int_{\mathbb{R}^2} \nabla'(\varphi_h(\xi - x')) \otimes \nabla'\overline{u}_h(\xi)\, d\xi = -\int_{\mathbb{R}^2} \nabla'\varphi_h(\xi) \otimes \nabla'\overline{u}_h(x' + \xi)\, d\xi,$$

where $\nabla'\varphi_h(\xi) = h^{-3}\nabla'\varphi(h^{-1}\xi)$Since the mollifier φ is radially symmetric, $\partial_{\xi_1}\varphi_h(\xi)$is an odd function of ξ1 and an even function of ξ2, and analogously for ∂ξ2φh(ξ)∂ξ2φh(ξ), hence we write

$$\partial_i\nabla'\widetilde{u}_h = -\int_{\{|\xi|\leq h,\, \xi_i > 0\}} \partial_i\varphi_h(\xi)\left(\nabla'\overline{u}_h(x' + \xi) - \nabla'\overline{u}_h(x' - \xi)\right) d\xi, \quad i = 1, 2.$$

Using the generalised Minkowski inequality once again we have, for i=1,2

$$\|\partial_i \nabla' \widetilde{u}_h\|_{L^2(\omega')} \leq \int_{\{|\xi| \leq h, \xi_i > 0\}} \left\|\partial_i \varphi_h(\xi)\left(\nabla' \overline{u}_h(x'+\xi) - \nabla' \overline{u}_h(x'-\xi)\right)\right\|_{L^2(\omega')} d\xi$$
$$\leq Ch^{-1} \max_{|\xi| \leq h} \left\|\nabla' \overline{u}_h(x'+\xi) - \nabla' \overline{u}_h(x'-\xi)\right\|_{L^2(\omega')} \leq C. \tag{13}$$

The last inequality is obtained analogously to (12).

Finally, we argue that $\nabla' \widetilde{u}_h$ is bounded pointwise independently of *h*:

$$\left|\nabla' \widetilde{u}_h(x')\right| = \left|\int_{\mathbb{R}^2} \varphi_h(\xi) \nabla' \overline{u}_h(\xi + x')\, d\xi\right| \leq \|\varphi_h(\xi)\|_{L^2(B_h)} \|\nabla' \overline{u}_h(\xi + x')\|_{L^2(B_h)}$$
$$\leq Ch^{-1} \left\|\nabla' \overline{u}_h(\xi + x') - R_h(\xi + x')\right\|_{L^2(B_h)} + Ch^{-1} \left\|R_h(\xi + x')\right\|_{L^2(B_h)} \leq C,$$

for all points x′∈ω and sufficiently small values of *h*.

As one can see from the above, $\nabla' \widetilde{u}_h$ is a “reasonably good” approximation of ∇′uh. It remains to approximate h-1∂3uh by a smooth vector function. Let $\widetilde{n}_h$ be the vector orthogonal to the surface $\widetilde{u}_h$ defined by

$$\widetilde{n}_h := \partial_1 \widetilde{u}_h \wedge \partial_2 \widetilde{u}_h.$$

We show that the L2-norm of n˜h−Rh,3 is at most of order *h*, where Rh,I stands for the *i*-th column of R_h. To this end we first transform the above expression for $\widetilde{n}_h$, as follows:

$$\begin{aligned}\widetilde{n}_h = \partial_1\widetilde{u}_h \wedge \partial_2\widetilde{u}_h &= \big(R_{h,1} + (\partial_1\widetilde{u}_h - R_{h,1})\big) \wedge \big(R_{h,2} + (\partial_2\widetilde{u}_h - R_{h,2})\big)\\ &= R_{h,3} + (\partial_1\widetilde{u}_h - R_{h,1}) \wedge R_{h,2} + R_{h,1} \wedge (\partial_2\widetilde{u}_h - R_{h,2})\\ &\quad + (\partial_1\widetilde{u}_h - R_{h,1}) \wedge (\partial_2\widetilde{u}_h - R_{h,2}).\end{aligned}$$

Since R_h and $\nabla'\widetilde{u}_h$ are L^∞-bounded, the L^2-norms of the last three terms are of order *h*, hence

$$\|\widetilde{n}_h - R_{h,3}\|_{L^2(\omega')} \leq Ch. \tag{14}$$

Additionally, it is clear from (13) and (10) that

$$\|\nabla'\widetilde{n}_h\|_{L^2(\omega')} \leq C. \tag{15}$$

Putting together the above estimates for $\widetilde{n}_h$ Theorems 2 and 1, we obtain the following result.

PROPOSITION 1

Let ω′ be a domain whose closure is contained in ω. Under the assumptions of Theorem 1 the following estimates hold:

$$\begin{aligned}&\left\|(\nabla'\widetilde{u}_h|\widetilde{n}_h) - \nabla_h u_h\right\|_{L^2(\Omega')} \leq Ch,\\ &\quad\left\|(\nabla'\widetilde{u}_h|\widetilde{n}_h) - R_h\right\|_{L^2(\omega')} \leq Ch.\end{aligned}$$

As is customary in periodic homogenisation, and due to the fact that our problem is of two-scale nature, we require a suitable notion of convergence which takes into account the microscopic behaviour, e.g. the so-called two-scale convergence, introduced by Nguetseng [19]. Out of several equivalent definitions of two-scale convergence, we will use the one based on periodic unfolding, see e.g. [5]. For convenience of the reader we give the definition of two-scale convergence that we use and some of its properties in Appendix 1. Next we prove a general statement about the two-scale convergence of a second gradient, without any reference to a specific problem. This assertion may be used in various contexts where two-scale convergence of a second gradient is required. In the present paper we make use of this statement in Theorems 4 and 9. In the remainder of this section we consider an arbitrary *d*-dimensional space, returning to the case d=2d=2 in Sect. 4 onwards.

THEOREM 3

(Second gradient and two-scale convergence) Let $\omega\subset R^d$ be a bounded domain and $Y:=[0,1)^d$. Assume that a sequence $v\varepsilon\in H2(\omega)$ is bounded in $H^2(\omega)H^2(\omega)$. Then up to a subsequence

$$\nabla^2 v_\varepsilon \overset{2}{\rightharpoonup} \nabla^2 v(x) + \nabla_y^2 \varphi(x, y), \tag{16}$$

where $v\in H^2(\omega)$ is the weak limit in $H^2(\omega)$ (hence, strong limit in $H1(\omega)$ of the sequence $v\varepsilon v\varepsilon$ and[2] $\varphi \in L^2(\omega; H^2_{\text{per}}(Y))$ with zero mean with respect to $y\in Y$.

Note that here $v\varepsilon v\varepsilon$ is a scalar function. The theorem directly applies to the vector case component-wise.

Proof

Following the idea of the proof of the two-scale convergence for the first gradients in [5] we introduce the following function of *x* and *y*:

$$\widetilde{v}_\varepsilon(x, y) := \varepsilon^{-1}\left(\varepsilon^{-1}\left(\mathcal{T}_\varepsilon(v_\varepsilon) - \mathcal{M}_Y(\mathcal{T}_\varepsilon(v_\varepsilon))\right) - y^c \cdot \mathcal{M}_Y(\mathcal{T}_\varepsilon(\nabla v_\varepsilon))\right), \tag{17}$$

where $T\varepsilon$ is the unfolding operator (see Appendix 1), $MY(f):=\int Yfdy$ is the operator of averaging with respect to the variable *y*, and $y^c := y-(1/2,\ldots,1/2)^T$. We calculate the first and second gradients of $\widetilde{v}_\varepsilon$ with respect to y:

$$\nabla_y \widetilde{v}_\varepsilon = \varepsilon^{-1}\left(\mathcal{T}_\varepsilon(\nabla v_\varepsilon) - \mathcal{M}_Y(\mathcal{T}_\varepsilon(\nabla v_\varepsilon))\right), \quad \nabla_y^2 \widetilde{v}_\varepsilon = \mathcal{T}_\varepsilon(\nabla^2 v_\varepsilon).$$

Notice that the function $\widetilde{v}_\varepsilon$ and its gradient $\nabla_y \widetilde{v}_\varepsilon$ have zero mean value in y. Hence, we can apply the Poincaré inequality:

$$\|\widetilde{v}_\varepsilon\|_{L^2(\omega\times Y)} \le C\|\nabla_y \widetilde{v}_\varepsilon\|_{L^2(\omega\times Y)} \le C\|\nabla_y^2 \widetilde{v}_\varepsilon\|_{L^2(\omega\times Y)} \le C,$$

where the last bound follows from the assumptions of the theorem. Hence, the sequence $\widetilde{v}_\varepsilon$ is bounded in $L^2(\omega;H^2(Y))$ and it converges weakly in this space, up to a subsequence:

$$\widetilde{v}_\varepsilon \rightharpoonup \widetilde{\varphi} \quad \text{weakly in } L^2(\omega; H^2(Y)).$$

Let a function $\widehat{\varphi} \in L^2(\omega; H^2(Y))$ be such that

$$\widetilde{\varphi} = \frac{1}{2} y^c \cdot (\nabla^2 v) y^c + \widehat{\varphi}. \tag{18}$$

Clearly, under this notation one has

$$\nabla_y^2 \widetilde{v}_\varepsilon \rightharpoonup \nabla^2 v(x) + \nabla_y^2 \widehat{\varphi}(x,y) \quad \text{weakly in } L^2(\omega\times Y),$$

which is equivalent to (16) upon setting

$$\varphi(x,y) := \widehat{\varphi}(x,y) - \int_Q \widehat{\varphi}(x,y)dy, \quad (x,y)\in\omega\times Y.$$

Now it only remains to prove the periodicity of φ with respect to y. To this end, let $\psi(x,y')$ be an arbitrary function from $C_0^\infty(\omega\times Y')$ where $y' := (y_1,\ldots,y_{d-1})^T$ and $Y' := [0,1)^{d-1}$. We write the difference between values of $\widetilde{v}_\varepsilon$ on the opposite faces of the cube Y corresponding to yd=1 and yd=0 (the proof for other components of y being exactly the same):

$$\begin{aligned}&\widetilde{v}_\varepsilon(x,y',1)-\widetilde{v}_\varepsilon(x,y',0)\\&=\varepsilon^{-2}\left(\mathcal{T}_\varepsilon(v_\varepsilon)(x,y',1)-\mathcal{T}_\varepsilon(v_\varepsilon)(x,y',0)\right)-\varepsilon^{-1}\mathcal{M}_Y(\mathcal{T}_\varepsilon(\partial_d v_\varepsilon))\\&=\varepsilon^{-2}\left(\mathcal{T}_\varepsilon(v_\varepsilon)(x,y',1)-\mathcal{T}_\varepsilon(v_\varepsilon)(x,y',0)-\mathcal{M}_Y(\partial_{y_d}\mathcal{T}_\varepsilon(v_\varepsilon))\right)\\&=\varepsilon^{-2}\left(\mathcal{T}_\varepsilon(v_\varepsilon)(x,y',1)-\mathcal{T}_\varepsilon(v_\varepsilon)(x,y',0)-\int_{Y'}\left(\mathcal{T}_\varepsilon(v_\varepsilon)(x,y',1)-\mathcal{T}_\varepsilon(v_\varepsilon)(x,y',0)\right)dy'\right).\end{aligned}$$

Notice that $\mathcal{T}_\varepsilon(v_\varepsilon)(x,y',1)=\mathcal{T}_\varepsilon(v_\varepsilon)(x+\varepsilon e_d,y',0)$, where eded is a unit vector in the direction of xd axis, hence we derive that

$$\begin{aligned}&\int_{\omega\times Y'}\left(\widetilde{v}_\varepsilon(x,y',1)-\widetilde{v}_\varepsilon(x,y',0)\right)\psi\,dxdy'\\&=\int_{\omega\times Y'}\varepsilon^{-2}\left(\mathcal{T}_\varepsilon(v_\varepsilon)(x,y',0)-\int_{Y'}\mathcal{T}_\varepsilon(v_\varepsilon)(x,y',0)dy'\right)\left(\psi(x-\varepsilon e_d,y')-\psi(x,y')\right)dxdy'\end{aligned}$$

(19)

The sequence $\varepsilon^{-1}(\psi(x-\varepsilon e_d,y')-\psi(x,y'))$ converges to $-\partial_d\psi$ strongly in $L^2(\omega\times Y')$. Let us denote

$$Z_\varepsilon(x,y):=\varepsilon^{-1}\left(\mathcal{T}_\varepsilon(v_\varepsilon)(x,y)-\int_{Y'}\mathcal{T}_\varepsilon(v_\varepsilon)(x,y)dy'\right).$$

Using a slight variation of the argument in Proposition 3.4 in [5] one can see that ZεZε converges strongly in $L^2(\omega;H^1(Y))$ to $(y^c)'\cdot\nabla'v$ as $\varepsilon\to 0$Hence, its trace

$$Z_\varepsilon(x,y',0)=\varepsilon^{-1}\left(\mathcal{T}_\varepsilon(v_\varepsilon)(x,y',0)-\int_{Y'}\mathcal{T}_\varepsilon(v_\varepsilon)(x,y',0)dy'\right)$$

converges to (yc)′·∇′v weakly in L2(ω×Y′) , see Proposition 2 below. Passing to the limit in (19) yields

$$\lim_{\varepsilon\to 0}\int_{\omega\times Y'}\left(\widetilde{v}_\varepsilon(x,y',1)-\widetilde{v}_\varepsilon(x,y',0)\right)\psi\,dxdy' = -\int_{\omega\times Y'}(y^c)'\cdot\nabla' v\,\partial_d\psi\,dxdy'. \tag{20}$$

We carry out the the same calculations for $z(x,y) := \frac{1}{2}\,y^c\cdot(\nabla^2 v)\,y^c$, which results in

$$z(x,y',1)-z(x,y',0) = (y^c)'\cdot\partial_d\nabla' v.$$

Multiplying the last expression by ψ and integrating by parts we get

$$\int_{\omega\times Y'}\left(z(x,y',1)-z(x,y',0)\right)\psi\,dxdy' = -\int_{\omega\times Y'}(y^c)'\cdot\nabla' v\,\partial_d\psi\,dxdy'.$$

Comparing the latter with (20) we conclude that φ(x,y) is periodic with respect to *y*.

Analogously we prove the periodicity of ∇yφ. Indeed, we have

$$\nabla_y\widetilde{v}_\varepsilon(x,y',1)-\nabla_y\widetilde{v}_\varepsilon(x,y',0) = \varepsilon^{-1}\left(\mathcal{T}_\varepsilon(\nabla v_\varepsilon)(x,y',1)-\mathcal{T}_\varepsilon(\nabla v_\varepsilon)(x,y',0)\right)$$

Note that by the assumptions of the theorem and properties of two-scale convergence (see Appendix 1) the sequence Tε(∇vε) converges to ∇v strongly in $L^2(\omega;H^1(Y))$. Hence, by Proposition 2 below the trace $\mathcal{T}_\varepsilon(\nabla v_\varepsilon)(x,y',0)$ converges to ∇v weakly in L2(ω×Y′), and therefore

$$\begin{aligned}
&\lim_{\varepsilon\to 0}\int_{\omega\times Y'}\left(\nabla_y\widetilde{v}_\varepsilon(x,y',1)-\nabla_y\widetilde{v}_\varepsilon(x,y',0)\right)\psi\,dxdy'\\
&= \lim_{\varepsilon\to 0}\int_{\omega\times Y'}\mathcal{T}_\varepsilon(\nabla v_\varepsilon)(x,y',0)\,\varepsilon^{-1}\left(\psi(x-\varepsilon e_d,y')-\psi(x,y')\right)dxdy'\\
&= -\int_{\omega\times Y'}\nabla v\,\partial_d\psi\,dxdy'.
\end{aligned}$$

Since $\nabla_y z(x, y', 1) - \nabla_y z(x, y', 0) = \partial_d \nabla v$ we have

$$\int_{\omega \times Y'} (\nabla_y z(x, y', 1) - \nabla_y z(x, y', 0))\, \psi \, dx dy' = - \int_{\omega \times Y'} \nabla v \, \partial_d \psi \, dx dy',$$

which implies the periodicity of ∇yφ∇yφ. Summarising all of the above we conclude that φ∈ $L^2(\omega; H^2_{\text{per}}(Y))$.

PROPOSITION 2

Suppose that a sequence fn(x,y)fn(x,y) converges to *f*(*x*, *y*) weakly in L2(ω;H1(Y))L2(ω;H1(Y)) and strongly in L2(ω×Y)L2(ω×Y). Then its trace on ω×∂Yω×∂Y converges to the trace of *f* weakly in L2(ω×∂Y)L2(ω×∂Y).

Proof

Without loss of generality we will assume that f=0 and prove the assertion for the part of the boundary corresponding to yn=0. We need to show that

$$\lim_{n \to \infty} \int_{\omega \times Y'} f_n(x, y', 0) \psi(x, y') \, dx dy' = 0, \tag{21}$$

for all $\psi \in C_0^\infty(\omega \times Y')$ Let us fix ψψ and consider it as a function defined on ω×Yω×Y via $\psi(x, y) := \psi(x, y')$ Let us define the function $h_n = h_n(y)$ as $\int_\omega f_n \psi \, dx$. It is easy to see that $h_n \in H^1(Y)$ $\|h_n\|_{L^2(Y)} \leq C \|f_n\|_{L^2(\omega \times Y)}$ $\|h_n\|_{H^1(Y)} \leq C \|f_n\|_{L^2(\omega; H^1(Y))}$. Hence hn→0 weakly in H1(Y) and strongly in L2(Y). By the trace theorem hn(y′,0)→0h strongly in L2(Y′). In particular,

$\int_{Y'} h_n(y',0)\,dy' \to 0$. But $h_n(y',0) = \int_\omega f_n(x,y',0)\psi(x,y')\,dx$, hence (21) follows.

Now we have all necessary tools to proceed to the ΓΓ-convergence statement.

4 LIMIT FUNCTIONAL AND ΓΓ-CONVERGENCE FOR THE "MODERATE" THICKNESS-TO-PERIOD RATIO E≫H≫E2

Our main result is establishing Γ-convergence of the rescaled functional Eh to the limit homogenised functional defined on isometric surfaces. Γ-convergence is a standard tool to describe the asymptotic behaviour of nonlinear functionals, introduced by De Giorgi [6]. The general definition of Γ-convergence is the following:

DEFINITION 1

Let FiFi, i=1,2,... and F∞ be some functionals defined on metric function spaces HiHi and H∞. Then we say that the sequence of functionals FiFi ΓΓ-converges to the functional F∞F∞,

$$F_\infty = \Gamma - \lim_{i\to\infty} F_i,$$

if the following two properties hold:

1. **LOWER BOUND.** For any sequence ui∈Hi converging in a certain sense to u∞∈H∞ one has

$$\liminf_{i\to\infty} F_i(u_i) \geq F_\infty(u_\infty).$$

2. **RECOVERY OF THE LOWER BOUND.** For any u∞∈H∞ with F∞(u∞)<∞ there exists a sequence ui∈Hiui∈Hi converging to u∞u∞ in a certain sense such that

limi→∞Fi(ui)=F∞(u∞).

In the above definition the notion of convergence depends on the underlying properties of the functionals and may be non-standard, in particular, because the spaces HiHi and H∞H∞ often do not coincide, as in the present paper, for example. Moreover, the notion of convergence in the second part of the definition may be different from the convergence in the first part. In this case the former must normally be "stronger" that the latter in order to facilitate the main purpose of the definition (nevertheless, this requirement is not always critical, e.g. see [2]). Under the assumption of equi-coercivity of the functionals FiFi on the underlying sequence of function spaces, the lower bound inequality ensures that the limit functional gives in the limit a lower bound for the values of the functionals FiFi. The recovery property implies that the lower bound F∞F∞ is the greatest lower bound. In particular, the definition implies that

$$\lim_{i\to\infty} (\inf F_i) = \inf F_\infty,$$

and, if a sequence uiui of "almost minimisers" of FiFi converges to u∞u∞, then the latter is a minimiser of F∞.

Remark 1

The ΓΓ-convergence statements proved in Theorems 4 and 9 below should be used in conjunction with the fact that the energy functionals involved are equi-coercive, see Theorem 1 above. In situations when the functionals are subject to appropriate boundary conditions, this implies convergence properties for the corresponding infima and minimising sequences.

In the present work we obtain different ΓΓ-limits for the regimes ε≫h≫ε2 and ε2≫h. In this section we present our results for the "moderate" regime ε≫h≫ε2, for which the limit homogenised functional is given by

$$E_{\text{hom}}^{\text{m}}(u) := \frac{1}{12} \int_{\omega} Q_{\text{hom}}^{\text{m}}(\text{II}) dx', \quad u \in H_{\text{iso}}^2(\omega),$$

where IIII is the matrix of the second fundamental form of *u*,

$$\text{II} := (\nabla' u)^{\top} \nabla' n, \quad n = \partial_1 u \wedge \partial_2 u,$$

and the homogenised stored-energy function is given by

$$Q_{\text{hom}}^{\text{m}}(\text{II}) := \min_{\psi \in H_{\text{per}}^2(Y)} \int_{Y} Q_2\left(y, \text{II} + \nabla_y^2 \psi(y)\right) dy, \tag{22}$$

where Q_2 is the quadratic form defined on 2×2 matrices and is obtained from the form Q_3 by minimising when reducing to two dimensions (notice that Q_3 vanishes on skew-symmetric matrices):

$$Q_2(y, A) := \min_{a \in \mathbb{R}^3} Q_3(y, A + e_3 \otimes a + a \otimes e_3) = \min_{a \in \mathbb{R}^3} Q_3(y, A + a \otimes e_3). \tag{23}$$

where we implicitly assume that the space $R^{2\times2}$ $R^{2\times2}$ matrices is naturally embedded in the space $R^{3\times3}$ by adding zero third column and zero third row, cf. [8].

THEOREM 4

Suppose that $\varepsilon \gg h \gg \varepsilon^2, i.e. \ \varepsilon^{-1}h = o(1)$ and $h^{-1}\varepsilon^2 = o(1) \ as \ h \to 0.$ Then the rescaled sequence of functionals $h^{-2}E_h$ Γ-converges to the limit functional $E^{\mathrm{m}}_{\mathrm{hom}}$ in the following sense.

1. **LOWER BOUND.** For every bounded bending energy sequence uh∈H¹(Ω) such that ∇′uh converges to ∇′u weakly in L2(Ω), $u \in H^2_{\mathrm{iso}}(\omega)$, the lower semicontinuity type inequality holds:

$$\liminf_{h\to 0} h^{-2}E_h(u_h) \geq E^{\mathrm{m}}_{\mathrm{hom}}(u).$$

2. **RECOVERY OF THE LOWER BOUND.** For every $u \in H^2_{\mathrm{iso}}(\omega)$there exists a sequence u_h^{rec} H1(Ω)uhrec∈H1(Ω) such that $\nabla' u_h^{\mathrm{rec}}$ converges to ∇′u strongly in L2(Ω) and

$$\lim_{h\to 0} h^{-2}E_h(u_h^{\mathrm{rec}}) = E^{\mathrm{m}}_{\mathrm{hom}}(u).$$

Proof

Lower bound. Since ∇hu is close to a rotation-valued function RhRh we can use the frame indifference property of the stored energy function *W* and its Taylor expansion near identity to linearise the functional. Roughly speaking, we will use the observation that

$$W(y, \nabla_h u_h) = W(y, R_h^\top \nabla_h u_h) = W\left(y, \mathcal{I}_3 + (R_h^\top \nabla_h u_h - \mathcal{I}_3)\right) \approx Q_3(y, hG_h). \tag{24}$$

where

$$G_h := h^{-1}\left(R_h^\top \nabla_h u_h(x) - \mathcal{I}_3\right).$$

However, the last asymptotic equality in (24) has to be made precise, since the smallness of the term hGhhGhis not pointwise. We also need to understand the structure of the (weak) two-scale limit of GhGh. To this end we will use a special representation of uhuh. Let ω′ω′ be a domain whose closure is contained in ω, and let Ω′=ω′×I For small enough *h* we can write uhuh in Ω′ in the form

$$u_h(x) = \widetilde{u}_h(x') + hx_3\widetilde{n}_h(x') + hz_h(x),$$

where u˜h is the mollified mid-surface of uhuh defined in the previous section and n˜h is the corresponding normal. Then one has

$$G_h = h^{-1}R_h^\top\left((\nabla'\widetilde{u}_h|\widetilde{n}_h) - R_h\right) + x_3R_h^\top(\nabla'\widetilde{n}_h|0) + R_h^\top\nabla_h z_h.$$

(25)

Notice that Gh is bounded in L2(Ω′) by Theorem 1. The first term on the right-hand side is bounded due to Proposition 1. The second term on the right-hand side is bounded by (15). Hence, the third term is also bounded.

Since we perform the dimension reduction when passing to the limit, we are interested only in the behaviour of the top left 2×2 block of the matrix GhGh. Let us use the following notation: if A∈R3×3, then Aˆ denotes the first two columns of *A*, and A¯ is obtained from *A* by omitting the third column and the third row. We have

$$\overline{G}_h = h^{-1}\widehat{R}_h^\top(\nabla'\widetilde{u}_h - \widehat{R}_h) + x_3\widehat{R}_h^\top\nabla'\widetilde{n}_h + \widehat{R}_h^\top\nabla'z_h. \quad (26)$$

The limit of the first and the third terms on the right-hand side of (26) is of no importance for us, apart from the fact that they do not depend on x3x3. This is obvious in the case of the first term. As for the third term, let $M_{Y\times I}(\mathcal{T}_\varepsilon(z_h)) := \int_{Y\times I} \mathcal{T}_\varepsilon(z_h)dydx_3$ and consider

$$w_h := \varepsilon^{-1}\left(\mathcal{T}_\varepsilon(z_h) - M_{Y\times I}(\mathcal{T}_\varepsilon(z_h))\right).$$

It is easy to see that $\nabla_y w_h = \mathcal{T}_\varepsilon(\nabla' z_h)$ and $\partial_3 w_h = \partial_3 \varepsilon^{-1}\mathcal{T}_\varepsilon(z_h) = \varepsilon^{-1} h \mathcal{T}_\varepsilon(h^{-1}\partial_3 z_h)$ Hence we have $\|\nabla_{y,x_3} w_h\|_{L^p(\Omega'\times Y)} \leq C$. Since $M_{Y\times I}(w_h) = 0$, we can use the Poincaré inequality $\|w_h\|_{L^p(\Omega'\times Y)} \leq C\|\nabla_{y,x_3} w_h\|_{L^p(\Omega'\times Y)}$, hence whwh is bounded in $L^p(\omega'; W^{1,p}(Y\times I))$. Therefore, up to a subsequence, we have

$$w_h \rightharpoonup w(x,y) \quad \text{weakly in } L^p\left(\omega'; W^{1,p}(Y\times I)\right).$$

On the other hand, we know that ∂3wh∂3wh vanishes in the limit, hence *w* is independent of x3x3:

$$w = w(x', y). \tag{27}$$

Therefore, we have

$$h^{-1}\widehat{R}_h^\top(\nabla'\widetilde{u}_h - \widehat{R}_h) + \widehat{R}_h^\top \nabla' z_h \overset{2}{\rightharpoonup} H(x', y) \tag{28}$$

for some $H \in L^2(\omega' \times Y)$, up to a subsequence.

Now we examine the expression $\widehat{R}_h^\top \nabla' \widetilde{n}_h$. The element (1, 1) of this matrix can be written as

$$\begin{aligned} R_{h,1}\cdot\partial_1\widetilde{n}_h &= R_{h,1}\cdot\partial_{11}\widetilde{u}_h \wedge \partial_2\widetilde{u}_h + R_{h,1}\cdot\partial_1\widetilde{u}_h \wedge \partial_{12}\widetilde{u}_h \\ &= \partial_{11}\widetilde{u}_h\cdot\partial_2\widetilde{u}_h \wedge R_{h,1} + \partial_{12}\widetilde{u}_h\cdot R_{h,1}\wedge\partial_1\widetilde{u}_h \\ &= -\partial_{11}\widetilde{u}_h\cdot R_{h,3} + \partial_{11}\widetilde{u}_h\cdot(\partial_2\widetilde{u}_h - R_{h,2})\wedge R_{h,1} + \partial_{12}\widetilde{u}_h\cdot R_{h,1}\wedge(\partial_1\widetilde{u}_h - R_{h,1}). \end{aligned}$$

where the vector product operations take priority over the inner products. Consider the second term in the last expression (the third term can be dealt in the same way). By Theorems 1 and 2 one has

$$\|\partial_{11}\widetilde{u}_h\|_{L^2(\omega')} \leq C, \quad \|\partial_2\widetilde{u}_h - R_{h,2}\|_{L^2(\omega')} \leq Ch' \quad \|\partial_2\widetilde{u}_h - R_{h,2}\|_{L^\infty(\omega')} \leq C.$$

On the one hand, the second term is bounded in L2(ω′), and so converges weakly two-scale to some function from L2(ω′×Y) up to a subsequence; on the other hand, for any C∞0(ω′×Y) test function we get zero in the limit. We conclude that the weak two-scale limits of the second and the third terms are zero. We can apply analogous argument to the other elements of $\widehat{R}_h^\top \nabla'\widetilde{n}_h$ Therefore we see that $\widehat{R}_h^\top \nabla'\widetilde{n}_h$ and $-\nabla'^2\widetilde{u}_h \cdot R_{h,3}$ have the same weak two-scale limit (the "dot" product gives a 2×2 matrix whose elements are −∂iju˜h·Rh,3). Applying Theorem 3 to u˜hu~h one has

$$\nabla'^2\widetilde{u}_h \xrightarrow{2} \nabla'^2 u(x') + \nabla_y^2\varphi(x', y)$$

up to a subsequence, for some $\varphi \in L^2(\Omega'; H^2_{\text{per}}(Y))$, whereas $R_{h,3}$ converges strongly to *n* due to Theorem 1. Hence,

$$\widehat{R}_h^\top \nabla'\widetilde{n}_h \xrightarrow{2} \left(-\nabla'^2 u + \nabla_y^2\varphi\right) \cdot n = \mathrm{II} + \nabla_y^2\psi, \tag{29}$$

where $\psi := -\varphi \cdot n \in L^2(\Omega'; H^2_{\text{per}}(Y))$ Putting together (28) and (29) we conclude that up to a subsequence

$$\overline{G}_h \xrightarrow{2} x_3\left(\mathrm{II}(x') + \nabla_y^2\psi(x', y)\right) + H(x', y). \tag{30}$$

We can proceed to the last step of the proof. In order to use the Taylor expansion of *W* we need to control G_h pointwise. Let χhχh be a characteristic function of the subset of Ω′ on which hG_h is relatively small, namely, χh:={10if |hGh|≤h1/2,otherwise.χh:={1if |hGh|≤h1/2,0otherwise.

(31)

It is easy to see that $\int_{\Omega'}(1-\chi_h)dx \to 0.$ Since W is non-negative we can restrict the integral to Ω′ and discard the set of points on which $|hG_h| > h^{1/2}$ by using χhχh:

$$\begin{aligned} h^{-2}E_h(u_h) &= h^{-2}\int\limits_{\Omega} W(\varepsilon^{-1}x', \nabla_h u_h)dx \geq h^{-2}\int\limits_{\Omega'} W(\varepsilon^{-1}x', \chi_h \nabla_h u_h)dx \\ &= h^{-2}\int\limits_{\Omega'} W\left(\varepsilon^{-1}x', \chi_h(\mathcal{I}_3 + hG_h)\right)dx = \int\limits_{\Omega'} Q_3(\varepsilon^{-1}x', \chi_h G_h)dx \\ &+ h^{-2}\int\limits_{\Omega'} r(\varepsilon^{-1}x', \chi_h hG_h)dx, \end{aligned}$$

(32)

where r is defined by (4). By the assumption (5) we have $\left|r(\varepsilon^{-1}x', \chi_h hG_h)\right| \leq g(\chi_h|hG_h|^2)$ for some function g(t)=o(t), , t→0. An elementary lemma below implies that3

$$h^{-2}\int\limits_{\Omega'} r\left(\varepsilon^{-1}x', \chi_h hG_h\right)dx \overset{h\to 0}{\longrightarrow} 0. \tag{33}$$

LEMMA 1

Suppose that $t^{-1}g(t) \to 0 \; as \; t \to 0,$ and let {ft}⊂L1(Ω′) be a family of non-negative functions such that $t^{-1}\int_{\Omega'} f_t$ is bounded with respect to t and $\| f_t \|_{L^\infty(\Omega')} \to 0$ as t→0. Then the convergence

$$t^{-1} \int_{\Omega'} g\left(f_t(x)\right) dx \stackrel{t \to 0}{\longrightarrow} 0$$

holds.

Proof

Notice that

$$\int_{\Omega'} \frac{g(f_t(x))}{t} dx = \int_{\Omega'} \frac{g(f_t(x))}{f_t(x)} \frac{f_t(x)}{t} dx \leq \left(\sup_{s \in (0, \|f_t\|_{L^\infty(\Omega')})} \frac{g(s)}{s} \right) \int_{\Omega'} \frac{f_t(x)}{t} dx \to 0.$$

Here it is implied, with some abuse of notation, that $(f_t(x))^{-1} g(f_t(x)) = 0$ whenever *ft*(x)=0 .

We apply Lemma 1 to the expression (33) by setting $t := h^2,\ f_t := \chi_h |hG_h|^2$ Notice that the particular choice of the cut-off threshold in (31) is not important: instead of h1/2 one can take any threshold that goes to zero slower than *h*.

Now we estimate the quadratic term in (32). By definition of the form Q2Q2 and the properties of the unfolding operator we have

$$\int_{\Omega'} Q_3(\varepsilon^{-1}x', \chi_h G_h) dx \geq \int_{\Omega'} Q_2(\varepsilon^{-1}x', \chi_h \overline{G}_h) dx \geq \int_{\Omega'} \int_{Y} Q_2\left(y, T_\varepsilon(\chi_h \overline{G}_h)\right) dy dx \qquad (34)$$

It is well known that convex integral functionals are lower semi-continuous with respect to the weak convergence, see e.g. [7]. Assume that we deal with a weakly converging (as in (30)) subsequence $\overline{G}_h$. Then passing to the limit and recalling that χhχh converges strongly to 1 we obtain the following inequality:

$$\begin{aligned}
&\liminf_{h\to 0} \int_{\Omega'}\int_{Y} Q_2\left(y, T_\varepsilon(\chi_h \overline{G}_h)\right)dydx \\
&\geq \int_{\Omega'}\int_{Y} Q_2\left(y, x_3(\mathrm{II}(x') + \nabla_y^2 \psi(x', y)) + H(x', y)\right)dydx \\
&= \frac{1}{12}\int_{\omega'}\int_{Y} Q_2(y, \mathrm{II} + \nabla_y^2 \psi)dydx' + \int_{\omega'}\int_{Y} Q_2(y, H)dydx' \\
&\geq \frac{1}{12}\int_{\omega'}\int_{Y} Q_2(y, \mathrm{II} + \nabla_y^2 \psi)dydx' \geq \frac{1}{12}\int_{\omega'}\int_{Y} Q_{\mathrm{hom}}^{\mathrm{m}}(\mathrm{II})dydx'.
\end{aligned} \tag{35}$$

Note that in the second step in the above we integrate with respect x3∈I, which causes the cross term of $x_3(\mathrm{II} + \nabla_y^2 \psi)$ and H in the quadratic form Q2Q2 to disappear. At the last step we just used the definition of $Q_{\mathrm{hom}}^{\mathrm{m}}$.

A simple argument by contradiction shows that the inequality

$$\liminf_{h\to 0} \int_{\Omega'}\int_{Y} Q_2\left(y, T_\varepsilon(\chi_h \overline{G}_h)\right) dydx \geq \frac{1}{12}\int_{\omega'}\int_{Y} Q_{\mathrm{hom}}^{\mathrm{m}}(\mathrm{II})dydx' \tag{36}$$

holds for the whole sequence $\overline{G}_h$. Putting together (32), (33), (34) and (36) and noting that ω′ is an arbitrary subset of ωω concludes the proof of the lower bound.

Remark 2

The above proof applies to the general case of h≪ε,h≪ε, up to the last inequality in (35). We will use this observation in our analysis of the lower bound in the case h≪ε2h≪ε2 in Sect. 7.

RECOVERY SEQUENCE. In order to build a recovery sequence one normally needs to ensure that the argument of the stored-energy density *W* is an L∞L∞-function, so that one can pass to the limit using the Taylor expansion (4). If this condition is not in place one would have to estimate *W* on the "bad" set where the argument might be uncontrollably large, which would require additional restrictions on *W*. For example in [8] the authors used a truncation result for Sobolev maps, which consists of changing a function on some subset to make it together with its gradient essentially bounded and estimating the measure of this set. In this paper we will use a result of Pakzad from [20] (which appeared after [8]). It states that the space of infinitely smooth isometric immersions from a convex set $\overline{\omega} \subset \mathbb{R}^2$ into $\mathbb{R}^3$ into R³ is strongly dense in the Sobolev space of isometric immersion $H^2_{\text{iso}}(\omega)$. Let $u \in H^2_{\text{iso}}(\omega) \cap C^\infty(\overline{\omega})$, and let IIII be the matrix of the corresponding second fundamental form. Let also ψ(x′,y) and d(x′,y) be correspondingly a scalar and a vector functions from $C^\infty(\overline{\omega} \times \overline{Y})$ periodic with respect to *y*.

Denote $\psi_\varepsilon := \psi(x', \varepsilon^{-1}x')$, $d_\varepsilon := d(x', \varepsilon^{-1}x')$ and consider the sequence

$$u_h = u - \varepsilon^2\psi_\varepsilon n + hx_3 n_\varepsilon + h^2\frac{x_3^2}{2}Rd_\varepsilon, \tag{37}$$

Where

$n := \partial_1 u \wedge \partial_2 u$, $n_\varepsilon := n + \varepsilon\partial_{y_1}\psi_\varepsilon\partial_1 u + \varepsilon\partial_{y_2}\psi_\varepsilon\partial_2 u$ and $R := (\nabla' u|n)$.
Calculating the rescaled gradient of uhuh we obtain

$$\nabla_h u_h = (\nabla' u - \varepsilon\nabla_y(\psi_\varepsilon n)|n_\varepsilon) + hx_3(\nabla' n + \nabla_y(\partial_{y_1}\psi_\varepsilon\partial_1 u + \partial_{y_2}\psi_\varepsilon\partial_2 u)|Rd_\varepsilon) + U_h,$$

where Uh denotes the remainder:

$$U_h := -\varepsilon^2(\nabla'(\psi_\varepsilon n)|0) + \varepsilon h x_3(\nabla'(\partial_{y_1}\psi_\varepsilon\partial_1 u + \partial_{y_2}\psi_\varepsilon\partial_2 u)|0)$$
$$+h^2\frac{x_3^2}{2}\left((\nabla' + \varepsilon^{-1}\nabla_y)(Rd_\varepsilon)|0\right).$$

Due to our assumptions we see that Uh=O(ε²)=o(h) uniformly on ωω. A simple calculation shows that

$$(\nabla'u - \varepsilon\nabla_y(\psi_\varepsilon n)|n_\varepsilon)^\top(\nabla'u - \varepsilon\nabla_y(\psi_\varepsilon n)|n_\varepsilon)$$
$$= \mathcal{I}_3 + \varepsilon^2\begin{pmatrix}(\partial_{y_1}\psi_\varepsilon)^2 & \partial_{y_1}\psi_\varepsilon\partial_{y_2}\psi_\varepsilon & 0\\ \partial_{y_1}\psi_\varepsilon\partial_{y_2}\psi_\varepsilon & (\partial_{y_2}\psi_\varepsilon)^2 & 0\\ 0 & 0 & |\nabla_y\psi_\varepsilon|^2\end{pmatrix} = \mathcal{I}_3 + o(h). \quad (38)$$

This implies that there exists an infinitely smooth function RhRh with values in SO(3) such that

$$R_h^\top(\nabla'u - \varepsilon\nabla_y(\psi_\varepsilon n)|n_\varepsilon) = \mathcal{I}_3 + o(h).$$

(It can be constructed explicitly via Gram–Schmidt orthonormalisation for the columns of the matrix $(\nabla'u - \varepsilon\nabla_y(\psi_\varepsilon n)|n_\varepsilon))$.

Further, consider the expression

$$G_h := h^{-1}(R_h^\top\nabla_h u_h - \mathcal{I}_3) = x_3 R_h^\top(\nabla' n + \nabla_y(\partial_{y_1}\psi_\varepsilon\partial_1 u + \partial_{y_2}\psi_\varepsilon\partial_2 u)|Rd_\varepsilon) + o(1).$$

Notice that RhRh converges to *R* uniformly on ω, $\nabla_y^2\psi_\varepsilon$ converges strongly two-scale to $\nabla_y^2\psi$, and dεdεconverges strongly two-scale to *d*. Hence, the convergence

$$G_h \xrightarrow{2} x_3(\mathrm{II} + \nabla_y^2\psi \mid d) \quad (39)$$

holds. Passing to the limit in

$$h^{-2}E_h(u_h) = h^{-2}\int_\Omega W(\varepsilon^{-1}x', \nabla_h u_h)dx$$
$$= \int_\Omega Q_3(\varepsilon^{-1}x', G_h)dx + h^{-2}\int_\Omega r(\varepsilon^{-1}x', hG_h)dx$$

we see that the remainder term vanishes since GhGh is uniformly bounded in L∞(ω), and for the quadratic term due to (39) we have

$$\int_\Omega Q_3(\varepsilon^{-1}x', G_h)dx = \int_{\Omega\times Y} Q_3(y, T_\varepsilon(G_h))dxdy + \int_{\Lambda_\varepsilon\times I} Q_3(\varepsilon^{-1}x', G_h)dx$$
$$\to \frac{1}{12}\int_{\omega\times Y} Q_3(y, (\mathrm{II} + \nabla_y^2\psi \mid d))dx'dy$$

(for the definition of ΛεΛε see Appendix 1). Thus

$$\lim_{h\to 0} h^{-2}E_h(u_h) = \frac{1}{12}\int_{\omega\times Y} Q_3\left(y, (\mathrm{II} + \nabla_y^2\psi \mid d)\right)dx'dy. \tag{40}$$

We conclude the proof with a standard argument showing that the above assumption of smoothness of *u*, ψ, is not restrictive. Let *u* be an arbitrary element of H2iso(ω) and take ψ=ψ(x′,y)∈L2 (ω; H2per (Y)) to be a solution to the minimisation problem in (22) corresponding to the matrix of the second fundamental form of *u*, that is

$$Q_{\mathrm{hom}}^{\mathrm{m}}(\mathrm{II}) = \int_Y Q_2(y, \mathrm{II} + \nabla_y^2\psi)dy.$$

Let also d(x′,y) be such that

$$Q_2(y, \mathrm{II} + \nabla_y^2\psi) = Q_3\left(y, (\mathrm{II} + \nabla_y^2\psi \mid d)\right).$$

There exist sequences of C∞C∞-functions ujuj, ψjψj and djdj converging to *u*, ψ and *d* in the corresponding function spaces such that

$$\left| \int_{\omega\times Y} Q_3\left(y, (\mathrm{II}_j + \nabla_y^2\psi_j \,|\, d_j)\right) dx'dy - \int_{\omega} Q^{\mathrm{m}}_{\mathrm{hom}}(\mathrm{II})dx' \right| \le \frac{1}{j}.$$

For each *j* define u(j)huh(j) via formula (37) with *u*, ψ and *d* replaced by ujuj, ψjψj and djdj. For each sequence u(j)hu the analogue of convergence (40) holds. It follows that there exists a subsequence hj→0 such that

$$\left| E_h(u^{(j)}_{h_j}) - \int_{\omega} Q^{\mathrm{m}}_{\mathrm{hom}}(\mathrm{II})dx' \right| \le \frac{2}{j}.$$

Hence urech:=u(j)hj is the required recovery sequence.

5 A MICROSCALE ISOMETRY CONSTRAINT ON THE LIMITS OF FINITE BENDING ENERGY SEQUENCES IN THE "SUPERCRITICAL" REGIME H≪E2

During the process of constructing a recovery sequence we have encountered a problem in the case when h≪ε2. Namely, introducing the corrector terms −ε2ψεn and ε∂y1ψε∂1u+ε∂y2ψε∂2u in (37) incurs an error of order ε2 (cf. (38)) which becomes of order h−1ε2 when scaled in the subsequent argument. This error does not vanish if h≪ε2 or h∼ε2. In order to understand the nature of the problem we write a formal asymptotic expansion for a recovery sequence, see Appendix 2. In the process of eliminating the errors of order ε2 and higher, it transpires that ψεψε needs to satisfy some solvability condition which is equivalent to the zero-determinant condition (41) below. As we show in the next theorem, this condition is intrinsic for the limits of finite bending energy sequences in the case h≪ε2 . This is due to the fact that II+∇2yψε approximates the matrix of the second fundamental form of the mollified mid-surface u˜h of the plate (defined in Sect. 3) on εε-scale. It appears that u˜hu∼h has to be almost isometric in the case h≪ε2 for the expression (∇′u˜h)T∇′u˜h−I2 to be of order $o(h)$. Noting that the matrix of the second fundamental form of an isometric surface has zero determinant suggests the condition det(II+∇2yψ)=0, where ψ is the two-scale limit of the sequence ψε.

THEOREM 5

(Zero determinant condition for the two-scale limit isometric surface) Let uhuh be a finite bending energy sequence, u˜h be the mollified mid-surface of the plate defined by (8) and ω′⋐ω. Assume that h≪ε2. Then the weak two-scale limit $\nabla'^2 u + \nabla_y^2 \varphi$ *of* $\nabla'^2 \tilde{u}_h$ $L^2(\omega' \times Y)$, $u \in H^2_{\mathrm{iso}}(\omega')$, $\varphi \in L^2(\omega'; H^2_{\mathrm{per}}(Y))$satisfies the "zero determinant" condition

$$\det(\mathrm{II} + \nabla_y^2 \psi) = (\mathrm{II}_{11} + \partial^2_{y_1 y_1} \psi)(\mathrm{II}_{22} + \partial^2_{y_2 y_2} \psi) - (\mathrm{II}_{12} + \partial^2_{y_1 y_2} \psi)^2 = 0, \tag{41}$$

where ψ:=−φ·n .

Proof

The proof is inspired by the general idea of the proof of the identity detII=0 for H2 isometric surfaces (see [20]). In our setting, however, the mollified mid-surface u˜hu∼h of the plate is not exactly isometric. This, together with the two-scale nature of our problem, requires in-depth analysis on the microscale.

The following identity is easily checked by a straightforward calculation (recall that u˜h is a C∞-function):

$$\det \nabla'^2 \widetilde{u}_h = \partial^2_{12}(\partial_1 \widetilde{u}_h \cdot \partial_2 \widetilde{u}_h) - \frac{1}{2}\left(\partial^2_{22}(\partial_1 \widetilde{u}_h)^2 + \partial^2_{11}(\partial_2 \widetilde{u}_h)^2\right).$$

Note that we can consider the third order tensor $\nabla^2 \widetilde{u}_h$ as a 2×2 matrix whose elements ∂iju˜h are vector-valued functions. Thus when we take its determinant we understand the multiplication of entries as the scalar product of vectors. We will use this convention henceforth. We can rewrite the above identity by applying the unfolding operator:

$$\mathcal{T}_\varepsilon(\det \nabla^2 \widetilde{u}_h) = \varepsilon^{-2}\left(\partial^2_{y_1 y_2} \mathcal{T}_\varepsilon(\partial_1 \widetilde{u}_h \cdot \partial_2 \widetilde{u}_h) - \frac{1}{2}\left(\partial^2_{y_2^2} \mathcal{T}_\varepsilon(\partial_1 \widetilde{u}_h)^2 + \partial^2_{y_1^2} \mathcal{T}_\varepsilon(\partial_2 \widetilde{u}_h)^2\right.\right. \quad (42)$$

We consider the left-hand side of the last identity as a scalar product of two "vectors $f_\varepsilon := (\mathcal{T}_\varepsilon(\partial^2_{11}\widetilde{u}_h), \mathcal{T}_\varepsilon(\partial^2_{12}\widetilde{u}_h))^\top$ and $g_\varepsilon := (\mathcal{T}_\varepsilon(\partial^2_{22}\widetilde{u}_h), -\mathcal{T}_\varepsilon(\partial^2_{12}\widetilde{u}_h))^\top$.

Since curlyfε=0 and divygε=0, we can apply the compensated compactness theorem, see e.g. [24], which implies that the left-hand side of (42) *-weak converges to the scalar product of their weak limits, that is

$$\lim_{h\to 0} \int_{\omega'} \int_Y \mathcal{T}_\varepsilon(\det \nabla'^2 \widetilde{u}_h)\eta\, dy\, dx' = \int_{\omega'} \int_Y \det(\nabla'^2 u + \nabla_y^2 \varphi)\eta\, dy\, dx'$$

for any $\eta \in C_0^\infty(\omega' \times Y)$. Note that x′ plays the role of a parameter, and does not affect our argument. Let us now consider the right-hand side of (42). Due to Proposition 1 we have $\partial_i \widetilde{u}_h = R_{h,i} + O(h),\ h \to 0,\ i = 1, 2,$ where $O(h)$ is understood in terms of the L2L2-norm. In particular, $\partial_1 \widetilde{u}_h \cdot \partial_2 \widetilde{u}_h = O(h), (\partial_i \widetilde{u}_h)^2 = 1 + O(h)$. Thus

$$\begin{aligned}
&\varepsilon^{-2}\left(\partial^2_{y_1 y_2} \mathcal{T}_\varepsilon(\partial_1 \widetilde{u}_h \cdot \partial_2 \widetilde{u}_h) - \frac{1}{2}\left(\partial^2_{y_2^2} \mathcal{T}_\varepsilon(\partial_1 \widetilde{u}_h)^2 + \partial^2_{y_1^2} \mathcal{T}_\varepsilon(\partial_2 \widetilde{u}_h)^2\right)\right) \\
&\quad = \varepsilon^{-2}\left(\partial^2_{y_1 y_2} O(h) - \frac{1}{2}\left(\partial^2_{y_2^2}(1 + O(h)) + \partial^2_{y_1^2}(1 + O(h))\right)\right) \\
&\quad = \partial^2_{y_1 y_2} O(h/\varepsilon^2) + \partial^2_{y_2^2} O(h/\varepsilon^2) + \partial^2_{y_1^2} O(h/\varepsilon^2).
\end{aligned}$$

(43)

Multiplying the last expression in (43) by an arbitrary function $\eta \in C_0^\infty(\omega' \times Y)$ and integrating by parts twice we see that it converges *-weakly to zero given that h/ε2=o(1), *i.e.*

$$\int_{\omega'} \int_Y \det\left(\nabla'^2 u(x') + \nabla_y^2 \varphi(x', y)\right) \eta dy dx' = 0.$$

This implies

$$\det\left(\nabla'^2 u(x') + \nabla_y^2 \varphi(x', y)\right) = 0. \tag{44}$$

At the last step of our proof we need one additional property of the two-scale limit φ(x′,y). We know that the second-order derivatives of u(x′) are parallel to the vector n(x′): one has

$$\begin{aligned}
\partial_{ij} u \cdot R_k &= \partial_i R_j \cdot R_k = \partial_j R_i \cdot R_k = \partial_i (R_j \cdot R_k) - R_j \cdot \partial_k R_i \\
&= -R_j \cdot \partial_k R_i = 0, \quad i, j, k \in \{1, 2\},
\end{aligned}$$

since at least two of the indices *i*, *j*, *k* coincide. Notice further that

$$\begin{aligned}-\partial_{ij}u\cdot n &= \partial_{1j}u\cdot(\partial_2 u\wedge\partial_i u)+\partial_{2j}u\cdot(\partial_i u\wedge\partial_1 u)\\ &= \partial_i u\cdot(\partial_{1j}u\wedge\partial_2 u)+\partial_i u\cdot(\partial_1 u\wedge\partial_{2j}u)\\ &= \partial_i u\cdot\partial_j(\partial_1 u\wedge\partial_2 u)=\partial_i u\cdot\partial_j n=\mathrm{II}_{ij},\quad i,j\in\{1,2\},\end{aligned}\tag{45}$$

hence ∂iju=−IIijn. In particular, one has det∇′2u=−detII. In order to derive (41) from (44) we need a similar property for φφ to be held. In fact, we require a much weaker property of ∂yiφ∂yiφ being orthogonal to RiRi for i=1,2.We establish it with the help of the following statement.

PROPOSITION 3

Let

$$z_\varepsilon(x,y):=\varepsilon^{-1}\left(\varepsilon^{-1}\left(\mathcal{T}_\varepsilon(\widetilde{u}_h)-\mathcal{M}_Y(\mathcal{T}_\varepsilon(\widetilde{u}_h))\right)-\mathcal{M}_Y(\mathcal{T}_\varepsilon(\nabla'\widetilde{u}_h))\,y^c\right),$$

(cf. (17)). Then

$$\|\partial_{y_i}z_\varepsilon\cdot\mathcal{T}_\varepsilon(R_{h,i})\|_{L^1(\omega'\times Y)}\to 0\quad as\ h\to 0,\quad i=1,2. \tag{46}$$

In the proof of this proposition we will need the following simple inequality, which is a kind of Poincaré inequality.

LEMMA 2

Let f∈H1(Y). Then

$$\int_{Y\times Y}|f(y)-f(\zeta)|^2dyd\zeta\le 2\int_Y|\nabla f(y)|^2dy.$$

Proof

Since the set of $C^\infty(Y)$ functions is dense in $H^1(Y)$ we only need to prove the lemma for $f \in C^\infty(Y)$. First we write down a pointwise estimate for $|f(y)-f(\zeta)|$:

$$
\begin{aligned}
|f(y)-f(\zeta)| &= \left| \int_{\zeta_1}^{y_1} \partial_1 f(t,\zeta_2)dt + \int_{\zeta_2}^{y_2} \partial_2 f(y_1,t)dt \right| \\
&\leq \int_0^1 \left(|\nabla f(t,\zeta_2)| + |\nabla f(y_1,t)|\right) dt.
\end{aligned}
$$

We finish the proof with

$$
\begin{aligned}
\int_{Y\times Y} |f(y)-f(\zeta)|^2 \, dy d\zeta &\leq \int_{Y\times Y} \int_0^1 \left(|\nabla f(t,\zeta_2)|^2 + |\nabla f(y_1,t)|^2\right) dt dy d\zeta \\
&= 2 \int_Y |\nabla f(y)|^2 \, dy.
\end{aligned}
$$

Proof (Proposition 3)

Taking the gradient of the function zεzε and using the properties of the unfolding operator we have

$$
\nabla_y z_\varepsilon = \varepsilon^{-1} \left(\mathcal{T}_\varepsilon(\nabla' \widetilde{u}_h) - \mathcal{M}_Y(\mathcal{T}_\varepsilon(\nabla' \widetilde{u}_h))\right).
$$

Let us write the L2-norm of the right-hand side (dropping the coefficient ε^{-1}) in a special form:

$$\begin{aligned}
&\|\mathcal{T}_\varepsilon(\nabla'\widetilde{u}_h) - \mathcal{M}_Y(\mathcal{T}_\varepsilon(\nabla'\widetilde{u}_h))\|^2_{L^2(\omega'\times Y)} \\
&= \int_{\omega'\times Y} \left| \mathcal{T}_\varepsilon(\nabla'\widetilde{u}_h) - \int_Y \mathcal{T}_\varepsilon(\nabla'\widetilde{u}_h)d\zeta \right|^2 dydx \\
&= \int_{\omega'\times Y} \left| \int_Y (\mathcal{T}_\varepsilon(\nabla'\widetilde{u}_h)(x,y) - \mathcal{T}_\varepsilon(\nabla'\widetilde{u}_h)(x,\zeta))d\zeta \right|^2 dydx \\
&\leq \int_{\omega'\times Y\times Y} \left| \mathcal{T}_\varepsilon(\nabla'\widetilde{u}_h)(x,y) - \mathcal{T}_\varepsilon(\nabla'\widetilde{u}_h)(x,\zeta) \right|^2 d\zeta dydx
\end{aligned} \tag{47}$$

Applying Lemma 2 to $\mathcal{T}_\varepsilon(\nabla'\widetilde{u}_h)$ we have

$$\begin{aligned}
&\int_{\omega'\times Y\times Y} \left| \mathcal{T}_\varepsilon(\nabla'\widetilde{u}_h)(x,y) - \mathcal{T}_\varepsilon(\nabla'\widetilde{u}_h)(x,\zeta) \right|^2 d\zeta dydx \leq 2\left\| \nabla_y \mathcal{T}_\varepsilon(\nabla'\widetilde{u}_h) \right\|^2_{L^2(\omega'\times Y)} \\
&= 2\varepsilon^2 \left\| \mathcal{T}_\varepsilon(\nabla'^2\widetilde{u}_h) \right\|^2_{L^2(\omega'\times Y)} \leq C\varepsilon^2.
\end{aligned} \tag{48}$$

where in the last inequality we used the property of boundedness of the second gradient of u˜h, see Theorem2. Due to Proposition 1 we can replace $\mathcal{T}_\varepsilon(\nabla'\widetilde{u}_h)$) in the last estimate by $\mathcal{T}_\varepsilon(\widehat{R}_h)$: :

$$\int_{\omega'\times Y\times Y} \left| \mathcal{T}_\varepsilon(\widehat{R}_h)(x,y) - \mathcal{T}_\varepsilon(\widehat{R}_h)(x,\zeta) \right|^2 d\zeta dydx \leq C\varepsilon^2. \tag{49}$$

Similarly to (47) we can write the expression in (46) as

$$\int_{\omega'\times Y\times Y} \varepsilon^{-1}\left(\mathcal{T}_\varepsilon(\partial_i\widetilde{u}_h)(x,y) - \mathcal{T}_\varepsilon(\partial_i\widetilde{u}_h)(x,\zeta)\right) \cdot \mathcal{T}_\varepsilon(R_{h,i})(x,y)d\zeta dydx.$$

As follows from Proposition 1, in order to prove the statement it is enough to show that

$$\varepsilon^{-1} \int\limits_{\omega' \times Y \times Y} \left(\mathcal{T}_\varepsilon(R_{h,i})(x,y) - \mathcal{T}_\varepsilon(R_{h,i})(x,\zeta) \right) \cdot \mathcal{T}_\varepsilon(R_{h,i})(x,y) d\zeta dy dx \to 0, \quad i = 1,2.$$

The expression under the sign of integral has the form (a−b)·a(a−b)·a at every point, where *a* and *b* are unit vectors. We use the following elementary identity, 2(a−b)·a=|a−b|[2], together with (49) to conclude the proof of the proposition:

$$\begin{aligned} &\varepsilon^{-1} \int\limits_{\omega' \times Y \times Y} \left(\mathcal{T}_\varepsilon(R_{h,i})(x,y) - \mathcal{T}_\varepsilon(R_{h,i})(x,\zeta) \right) \cdot \mathcal{T}_\varepsilon(R_{h,i})(x,y) d\zeta dy dx \\ &= \frac{1}{2}\varepsilon^{-1} \int\limits_{\omega' \times Y \times Y} \left| \mathcal{T}_\varepsilon(R_{h,i})(x,y) - \mathcal{T}_\varepsilon(R_{h,i})(x,\zeta) \right|^2 d\zeta dy dx \le C\varepsilon. \end{aligned}$$

Theorem 3 implies that the sequence zεzε converges weakly in L[2](ω′;H[2](Y)) to the function (cf. (18))

$$z = \frac{1}{2}(y_1^c)^2 \partial_1^2 u + y_1^c y_2^c \partial_1 \partial_2 u + \frac{1}{2}(y_2^c)^2 \partial_2^2 u + \varphi.$$

Since Tε(Rh) converges to *R* strongly in L2(ω′×Y), we have

$$\partial_{y_i} z_\varepsilon \cdot \mathcal{T}_\varepsilon(R_{h,i}) \rightharpoonup \partial_{y_i} z \cdot R_i,$$

*-weakly, i.e. with test functions taken from C∞0(ω′×Y) (In fact, it is possible to show weak convergence in L2(ω′×Y), however, *-weak convergence is sufficient for our purposes). Further, since the second-order derivatives of *u* are parallel to the vector *b*, we have

$$\partial_{y_i} z \cdot R_i = \partial_{y_i} \varphi \cdot R_i.$$

Lastly, it follows from Proposition 3 that the *-weak limit of ∂yizε·Tε(Rh,i) equals zero, hence

$$\partial_{y_i} \varphi \cdot R_i = 0, \quad i = 1,2.$$

The above orthogonality property implies that the derivatives of φφ can be represented in the form $\partial_{y_1}\varphi = (\partial_{y_1}\varphi \cdot R_2)R_2 + (\partial_{y_1}\varphi \cdot n)n,\ \partial_{y_2}\varphi = (\partial_{y_2}\varphi \cdot R_1)R_1 + (\partial_{y_2}\varphi \cdot n)n.$ Differentiating further we get the following identities for the second derivatives of φφ (in particular we see that the mixed derivative is orthogonal to both R1 and R2):

$$\begin{aligned}
\partial_{y_1}^2\varphi &= (\partial_{y_1}^2\varphi \cdot R_2)R_2 + (\partial_{y_1}^2\varphi \cdot n)n,\\
\partial_{y_2}^2\varphi &= (\partial_{y_2}^2\varphi \cdot R_1)R_1 + (\partial_{y_2}^2\varphi \cdot n)n,\\
\partial_{y_1}\partial_{y_2}\varphi &= (\partial_{y_1}\partial_{y_2}\varphi \cdot n)n.
\end{aligned}$$

Substituting the latter and ∇′2u=−IIn into (44) and recalling that ψ=−φ·n we obtain (41).

6 THE STRUCTURE OF ISOMETRIC IMMERSIONS

The idea of the construction of a recovery sequence for the supercritical regime h≪ε2, which is performed in the next section, relies on the properties of isometric surfaces from H2iso(ω). There has recently been a number of works devoted to this topic, see [12, 13, 16, 20]. The results of Sect. 5 (as discussed at the beginning of Sect. 7) imply a discontinuous dependence of the homogenised energy density function on the direction of the bending. This observation drastically reduces the number of manipulations one can perform with a surface in order to construct a recovery sequence. In particular, we can not use the result on approximation of H2iso(ω) -surfaces by infinitely smooth ones because this approximation does not preserve the orientation of the bending directions. In our paper we will use mostly the results of [20], which we review below. We then give an example of an isometric surface for which the set of rational directions is irregular (cf. Theorem 8), and prove a statement on regularisation of isometric surfaces required for the construction of a recovery sequence.

6.1 Auxiliary observations

For any u∈H2iso(ω) the domain ωω can be partitioned into "bodies" and "arms", according to the following definition.

DEFINITION 2

A **BODY** is a maximal subdomain of ωω on which *u* is affine and whose boundary contains at least three segments inside ωω.

A differentiable curve Γ:[0,l]→ω is referred to as a **LEADING CURVE** if *u* is affine when restricted to the lines normal to Γ([0,l]), which we refer to as **LEADING SEGMENTS**. In the approximation procedure below we also use the term "leading curve" for the curve γ:=(u∘Γ) on the surface.

For a given leading curve Γ, **GENERATRICES** are the lines tangent to the surface *u* at the points of γ and orthogonal to the latter.

An **ARM** ω[Γ] is a subdomain of ωω covered by leading segments corresponding to a leading curve Γ (See Remark 4 below).

Remark 3

Note that a maximal subdomain of ωω on which *u* is affine and whose boundary contains at most two segments inside ωω is always an arm or part of an arm.

Remark 4

Suppose that Γ is a leading curve parametrised by its arclength, and denote by T:=Γ′ the tangent vector and by N:=(−T2,T1)T the unit vector orthogonal to Γ. Then the corresponding arm is given by

$$\omega[\Gamma] = \{\Gamma(t) + sN(t) : s \in \mathbb{R}, t \in [0, l]\} \cap \omega. \qquad (50)$$

For each t∈[0,l] the set {Γ(t)+sN(t):s∈R}∩ω is the corresponding leading segment.

Since leading curves are not uniquely defined, it is more reasonable to define the vector field *N* on the whole ω[Γ] as a continuous vector field of unit vectors parallel to leading segments, and the vector field *T* as the continuous unit vector field orthogonal to *N*. In particular, we have T=Γ′ for any leading curve Γ . In what follows we consider *N* and *T* either as vector fields N(x′), T(x′) or the normal *N*(*t*) and the tangent *T*(*t*) of the curve Γ(t), depending on the context. We will say that the value of *T* (equivalently, *N*) is **RATIONAL** if the ratio of its components is rational. We will also say in this situation that the corresponding leading segment is rational.

Remark 5

It can be shown that:

(a)For any given body or arm, the part of its boundary that is situated in the interior of ωω is a union of straight segments whose vertices belong to ∂ω;∂ω;

(b)As follows from [20], all leading curves are in the class C1,1loc.Cloc1,1. A combination of this observation with the result of [16] establishing that H2iso⊂C1(ω)Hiso2⊂C1(ω) implies the fact that leading curves γγ on the surface *u* are in the class C1C1.

Since the leading segments do not intersect in ωω, the vector field *N* (and, hence, *T*) is locally Lipschitz in ωω. In is not difficult to prove, using a geometric argument, the following lemma.

LEMMA 3

The local Lipschitz constant

$$C_L(x_0') := \limsup_{x'\to x_0'} \frac{|N(x') - N(x_0')|}{|x' - x_0'|} = \limsup_{x'\to x_0'} \frac{|T(x') - T(x_0')|}{|x' - x_0'|}, \qquad x_0' \in \omega,$$

satisfies the estimate

$$C_L(x_0') \le \mathrm{dist}^{-1}(x_0', \partial\omega).$$

This lemma implies, in particular, local Lipschitz continuous differentiability of leading curves.

Note that for any leading curve Γ, unit tangent vectors of γ are also tangent to the surface $u:R^2\to R^3$. Therefore, the surface *u* is reconstructed on each arm from the knowledge of the corresponding leading curve and the non-vanishing principal curvature. Indeed, denote by κΓ the curvature of Γ, so that Γ″=T′=κΓN . For a given γ, we define a Darboux frame (τ,ν,n) by the tangent τ=γ′ to the curve γ, the unit vector ν=(∇′u)N tangent to *u* and orthogonal to τ, and n=τ∧ν. This Darboux frame satisfies the system of equations (see e.g. [22, p. 277], [20])

$$\begin{pmatrix} \tau' \\ \nu' \\ n' \end{pmatrix} = \begin{pmatrix} 0 & \kappa_\Gamma & \kappa \\ -\kappa_\Gamma & 0 & 0 \\ -\kappa & 0 & 0 \end{pmatrix} \begin{pmatrix} \tau \\ \nu \\ n \end{pmatrix} \qquad (51)$$

which has a unique solution subject to appropriate initial values. Given γ(t) and ν(t) we can write u|ω[Γ] as

$$u = \gamma(t) + s\nu(t), \tag{52}$$

where *s* parametrises the generatrix corresponding to the value *t* on the curve Γ.Γ.

Remark 6

There is freedom in the choice of the sign of κΓ and the direction of the vector field *N*. We make this choice in such a way that the normal *n* in the above Darboux frame coincides with the normal *n* defined as ∂1u∧∂2uu in the discussion preceding the current section.

Let us write down the formulae for for the gradients ∇′u,∇′2u and the matrix IIII in terms of the leading directions and curvatures. Since x′ is related to (*t*, *s*) via x′(t,s)=Γ(t)+sN(t). We have

$$\frac{\partial(x_1, x_2)}{\partial(t, s)} = (\Gamma' + sN', N) = \begin{pmatrix} T_1(1 - s\kappa_\Gamma) & -T_2 \\ T_2(1 - s\kappa_\Gamma) & T_1 \end{pmatrix}.$$

and therefore

$$\frac{\partial(t, s)}{\partial(x_1, x_2)} = \begin{pmatrix} T_1(1 - s\kappa_\Gamma)^{-1} & T_2(1 - s\kappa_\Gamma)^{-1} \\ -T_2 & T_1 \end{pmatrix}.$$

The above expression is well defined since 1−sκΓ is positive everywhere on ω[Γ], cf. Lemma 3. Then we can write the gradient and the second gradient of *u* in terms of the geometry of the curves γ and Γ:

$$\nabla' u = (T_1\tau - T_2\nu,\ T_2\tau + T_1\nu),$$

$$\nabla'^2 u = \frac{\kappa}{1 - s\kappa_\Gamma} \begin{pmatrix} T_1^2 n & T_1 T_2 n \\ T_1 T_2 n & T_2^2 n \end{pmatrix}. \tag{53}$$

Notice that (cf. (45))

$$\mathrm{II} = -\frac{\kappa}{1 - s\kappa_\Gamma}\begin{pmatrix} T_1^2 & T_1 T_2 \\ T_1 T_2 & T_2^2 \end{pmatrix}. \tag{54}$$

Let us introduce the following notation

$$\widetilde{\mathrm{II}} = \begin{pmatrix} T_1^2 & T_1 T_2 \\ T_1 T_2 & T_2^2 \end{pmatrix}. \tag{55}$$

As will be shown in the next section, the homogenised stored energy density QschomQhomsc in the supercritical regime is in general a discontinuous function of its argument IIII. More precisely, it is continuous with respect to the coefficient −κ(1−sκΓ)−1, but discontinuous with respect to II˜ (the direction *T*). This suggests that one can modify the curvature κκ in a continuous way (*i.e.* without changing much the homogenised energy), however, a perturbation of the vector field *T* may lead to uncontrollable changes in the homogenised energy.

6.2 Regularisation of irregular surfaces.

The results of [20] and [16] assert some regularity of isometric immersions u∈H2iso(ω), namely, u∈C1(ω) and its associated leading curves in ωω have locally Lipschitz derivative. Nevertheless, isometric immersions can be quite irregular for the purpose of this paper. To illustrate this, we give an example of a surface for which the set of "rational" directions for the corresponding leading curve is the union of a fat Cantor set and a set of zero measure.

EXAMPLE. Let us recall the construction of a fat Cantor set. At the first step we remove from the middle of [0, 1] an open interval B1 of length 1 / 4 and denote the remaining set A1. At the next step from the middle of each the two subintervals of A1 we remove open intervals of length 1 / 16 each, the remaining set is denoted by A2 and its complement by B2 and so on: at the *n*-th step we remove 2n−1 disjoint intervals of length 1/22n, whose union is a set of measure 1/2n+1. Thus we have λ1(A)=λ1(B)=12, where A:=∩nAn, B:=∪nBn. By λdλd we denote the Lebesgue measure in RdRd. The fat Cantor set *A* is closed and is nowhere dense (i.e. contains no interior points). For each *n* we define a leading curve via its unit tangent vector Tn(t)Tn(t) given as follows. Let ϕϕ be a function that is continuously differentiable on the interval [0, 1], has zero mean and is such that $\varphi(0) = \varphi(1) = \varphi'(0) = \varphi'(1) = 0$, e.g. $\varphi(t) = \sin(2\pi t)^3$, and set $A0 = [0,1]$, $B0 = \emptyset$, $T0 := (1,0)^\mathsf{T}$. For each $n = 1,2,\ldots,$, we define $Tn := Tn-1$ on the set $[0, 1]\backslash(Bn \backslash Bn-1)$, and on each component interval In,k of the set $Bn \backslash Bn-1$ we define

$$T_n := \left(\sqrt{1-\phi_{n,k}^2},\ \phi_{n,k}\right)^\mathsf{T}, \quad k = 1, 2, \ldots,$$

where $\varphi n,k = \beta n\, \varphi(2^{2n}t - tn,k)$ is the scaled (by a coefficient βn, $|\beta n| < 1$, and the length of the interval $1/2^{2n}$) and appropriately translated (by the left end tn,k of the interval In,k) version of φ, which “fits” into this interval. We then integrate Tn to obtain the curves $\hat{W}n$:

$$\Gamma_n(t) := \int_0^t T_n, \quad t \in [0, 1].$$

Note that at each step of this procedure we "redefine" the vector T_n only on the set $B_n \setminus B_{n-1}$. We take $\beta_n = \beta/2^{2n}$, where $0 < \beta < 4$ max $|\varphi|^{-1}$ is a constant that will be specified later. So far we have a sequence of curves $\hat{W}_n$ with a bounded second derivative[4]: $\Gamma_n'' \le C\beta$. $|\Gamma''| \le C\beta$. It has a subsequence that converges strongly in H1(0,1)H1(0,1) to a curve Γ∈H2(0,1) with |Γ″|≤Cβ|Γ″|≤Cβ. In particular, one has $\hat{W}' = (1,0)^T$ on the fat Cantor set defined above, and the subset of B on which $\hat{W}'$ takes rational directions has zero measure. Let $\omega := (0,l) \times (0,1)$, where $l < 1$ is defined by the right end of $\hat{W}$, namely it is the first coordinate of the point $\hat{W}(1)$. If we choose β so that $|\hat{W}''| \le C\beta \le (\text{diam}(\omega))^{-1}$, the lines normal to $\hat{W}$ do not intersect within ω. (Clearly, one has diam $(\omega) \le \sqrt{2}$, so the choice $\beta \le 1/(C\sqrt{2})$) satisfies the required condition.) Lastly, we can construct an isometric surface u via the procedure described above [see formulae (51, 52)] with some "reasonable" curvature κκ, e.g. κ≡1κ≡1. For thus constructed u the domain ωω is the only arm and ΓΓ is its leading curve in the sense of Definition 2.

The above example illustrates that for a leading curve Γ(t)Γ(t) the set on which $T(t)$ has rational directions can not in general be represented as a union of intervals (up to a null set). In what follows we argue that one can replace an arbitrary isometric surface u by a "sufficiently regular" isometric surface, in order to obtain an explicit construction of a recovery sequence. Note that one can not simply approximate u by an infinitely smooth isometric surface, since Qschom,Qhomsc, in general, is an everywhere discontinuous function of its argument. Thus the matrix II˜II~, see (55), of the replacement surface should be different from the one of u only on a relatively small set. First we need to recall some properties of isometric surfaces. We reformulate and complement some results of [20] (Lemmas 3.8 and 3.9) in the following statement.

THEOREM 6

Let $u \in H^2$ (ω). *For any* δ > 0 *there exists* $v \in H^2$ (ω) *such that*

$$\lambda_2(\{x' : u(x') \neq v(x')\}) + \|u - v\|_{H^2(\omega)} \leq \delta,$$

the surface *v* has a finite number of bodies and arms, each maximal connected region with non-zero mean curvature lies between two regions where *v* is affine, the corresponding leading curves are Lipschitz differentiable and do not have common points with ∂ω∂ω.

The above properties of *v* are partly derived from the proofs of Lemmas 3.8 and 3.9 in [20]. Next, we briefly elucidate them. Every maximal connected region (we denote it by ΣΣ) where *u* has non-zero mean curvature is attached to either one or two regions where *u* is affine. If ΣΣ is situated between two regions where *u* is affine, then the lengths of the leading segments that cover ΣΣ are bounded below by a positive constant. Lemma 3 implies that on any closed subset of ωω the vector field *T* is Lipschitz continuous. In particular, we can cover ΣΣ by a finite number of leading curves so that they do not have common points with the boundary of ωω, and hence have Lipschitz continuous derivative. If ΣΣ has just one "affine" region attached to it or Σ=ωΣ=ω, then it may happen that the infimum of the lengths of leading segments covering ΣΣ is zero. Then one can not control the Lipschitz constant on the sequence of segments whose lengths tend to zero (such sequence will converge to a point on the boundary of ωω). In this case we can replace *u* in a small region near the boundary point where such degeneration of leading segments occurs by an affine function, making sure that the new surface is still an element of H2iso(ω)Hiso2(ω). Then we arrive at the situation discussed previously.

As was mentioned at the beginning of this section, we aim at the construction of recovery sequences for the homogenised functional $E_{\text{hom}}^{\text{sc}}$ in the supercritical case h≪ε2, given by (67).

Theorem 6 implies, in particular, that for δ>0 there exists v∈H2iso(ω) with a finite number of bodies and arms that depends on δδso that

$$\left|E_{\mathrm{hom}}^{\mathrm{sc}}(u) - E_{\mathrm{hom}}^{\mathrm{sc}}(v)\right| \leq \delta.$$

The next theorem ensures that we can replace $u \in H_{\mathrm{iso}}^{2}(\omega)$ with an isometric surface such that the set on which its leading segments are rational can be represented as a union of finite number of arms and a set of arbitrary small measure.

THEOREM 7

Let u∈H2iso(ω) be such that the subset ωr(u) of ω covered by rational leading segments has positive measure. Then for each δ>0 there exists a surface uδ∈H2iso(ω) such that:

- The surface uδuδ has a finite number of bodies and arms, which we generically denote by ω[Γ~];
- The subset ωr(uδ) of ω covered by rational leading segments can be represented as

$$\omega_r(u_\delta) = \left(\bigcup_{\tilde{\Gamma}_r} \omega[\tilde{\Gamma}_r]\right) \cup S_\delta,$$

- where each arm ω[Γ~r] consists of parallel rational segments (in particular, each leading curve Γ~r is a straight segment and u(ω[Γ~r]) is cylindrical);

- The set Sδ is measurable;

- The estimate

$$\lambda_2(S_\delta) + \|u - u_\delta\|_{H^2(\omega)} + |E^{sc}_{hom}(u) - E^{sc}_{hom}(u_\delta)| < \delta$$

(56)

holds.

Moreover, uδ can be chosen in such a way that on each arm the corresponding curvature κκ (cf. (51), (54)) is infinitely smooth and $II_\delta \in L^\infty(\omega)$, where the latter is the matrix of the second fundamental form of uδ.

Remark 7

The conditions that the curvature κκ is smooth and that $II_\delta \in L^\infty(\omega)$ are required for the construction of a recovery sequence. In particular, we take advantage of these facts when using the Taylor expansion formula for *W*, see (4).

Proof

Proof In view of Theorem 6 we may assume without loss of generality that *u* has a finite number of bodies and arms from the outset. Let us consider an arm ω[Ŵ] with the corre- sponding leading curve Ŵ(*t*), *t* ∈ [0,*l*], and assume that the subset *K* = *K*(Ŵ) of [0,*l*] on which the directions of *T* = *T*(*t*) are rational has positive measure λ1(*K*) > 0. Since the set of rational directions is countable, we can order its elements in such a way that λ1(*K*1) ≥ λ1(*K*2) ≥ .. ., where *Ki* is the preimage of a particular rational direction *Ti*, *i* = 1,2,3,..., so that *K* = ∪*i* *Ki*. We take an arbitrary η > 0 and let *m* = *m*(η) be such that for Kη:=∪mi=1Ki we have λ1(K\Kη)≤η. Note that all Ki are closed, in particular, the distance between any two of these sets is positive. This implies that we can remove a finite number of open intervals $\tilde{L}_i \subset [0,l] \setminus K_\eta, i=1\ldots,k$, so that the remaining set $L_\eta := [0,l] \setminus (\cup_{i=1}^k \tilde{L}_i)$ (we may also remove the points 0 and l from L_η if they are isolated) possesses the following properties:

- The estimate $\lambda_1(L_\eta \setminus K_\eta) < \eta$ holds.
- The set L_η is the union of a finite number of mutually disjoint closed intervals L_j,

  $j = 1, \ldots, n$.
- For each j one has $L_j \cap K_\eta = L_j \cap K_i$ for some $i \in \{1,2,\ldots,k\}$ and both ends of the interval L_j belong to K_i.

Let us consider a particular interval $L_j = [a_j, b_j]$. The part of $\omega[\hat{W}]$ covered by the restriction of $\hat{W}$ to L_j, which we denote by $\omega[\hat{W}|L_j]$, is contained between two parallel segments $\hat{W}(a_j) + s N_i$ and $\hat{W}(b_j) + s N_i$ corresponding to the rational direction T_i. In fact, loosely speaking, $\omega[\hat{W}|L_j]$ is mostly covered by the leading segments which are parallel to N_i. On the remainder of the set, corresponding to the restriction $\hat{W}|L_j \setminus K_i$, where the leading segments are not parallel to N_i, we replace them with new leading segments that are parallel to N_i. Notice that the difference between $N(t)$ and N_i on the set $L_j \setminus K_i$ is not greater than $C\eta$, where C is the Lipschitz constant of $N(t)$. We can write

$$\omega[\Gamma|_{L_j}] = \{\Gamma(t) + sN_i,\ s \in \mathbb{R},\ t \in L_j\} \cap \omega.$$

We carry out this procedure for each interval L_j. In this way we change the leading segments covering $\omega[\hat{W}]$ only on a small set $\omega_\eta[\hat{W}]$. We estimate the measure of this set as follows:

$$\lambda_2\left(\omega_\eta[\Gamma]\right) \leq \lambda_1(L_\eta \setminus K_\eta)\mathrm{diam}(\omega) \leq \eta\, \mathrm{diam}(\omega). \quad (57)$$

Let us denote by $N\eta$ and $T\eta$ the vector fields corresponding to the new set of leading segments. We have

$$N_\eta = N,\ T_\eta = T \quad \text{in } \omega[\Gamma]\backslash\omega_\eta[\Gamma].$$
$$\|N_\eta - N\|_{L^\infty(\omega_\eta[\Gamma])} \to 0,\ \|T_\eta - T\|_{L^\infty(\omega_\eta[\Gamma])} \to 0 \quad \text{as } \eta \to 0.$$

(58)

We denote by $S^{\hat{W}}$ the union of all rational leading segments that do not belong to the set $j = 1$ $\omega[\hat{W}|L_j]$ and estimate its measure as follows:

$$\lambda_2\left(S_\eta^\Gamma\right) \leq \lambda_1(K\backslash K_\eta)\text{diam}(\omega) \leq \eta\,\text{diam}(\omega). \tag{59}$$

The idea behind the rest of the proof is as follows. For each arm arm we build a family of leading curves $\hat{W}\eta$ and isometric surfaces $U\eta$, which will be shown to converge to the surface u uniformly as $\eta \to 0$. We assemble the approximating parts $U\eta$ on all arms and bodies in a continuous fashion. We then choose a sufficiently small value of η so that the resulting surface denoted by $u\delta$ satisfies the inequality (56).

Proceeding with this plan, we define $\hat{W}\eta$ as the maximal solution in $\omega[\hat{W}]$ to the following ODE:

$$\Gamma'_\eta(t) = T_\eta\left(\Gamma_\eta(t)\right), \quad \Gamma_\eta(0) = \Gamma(0), \tag{60}$$

where $T\eta$ is the field of unit vectors orthogonal to the modified set of leading segments of $\omega[\hat{W}]$. Notice that $\hat{W}\eta$ uniformly tends to $\hat{W}$ as $\eta \to 0$. In particular, for small enough values of η, the new curve will cover the whole of $\omega[\hat{W}]$, i.e. one has $\omega[\hat{W}] = \omega[\hat{W}\eta]$. We denote the domain of $\hat{W}\eta$ by $[0, l\eta]$, noting that in general $l\eta = l$. Both T an $T\eta$ are Lipschitz continuous on any compact subset of ω, and the measure of the set $\omega\eta[\hat{W}]$ on which $T = T\eta$ is proportional to η. This implies that on $\omega[\hat{W}\eta]$ one has a pointwise estimate $|T(t) - T\eta(t)| \leq C\eta$, where C is proportional to the Lipschitz constant. It follows that the solution of (60) converges to $\hat{W}$ uniformly as $\eta \rightarrow 0$. For each leading segment of $\omega[\hat{W}]$ there are unique $t \in [0,l]$ and $\hat{t} \in [0, l\eta]$ such that $\hat{W}(t)$ and $\hat{W}\eta(\hat{t})$ belong to this segment. Let us denote by $\zeta\eta$ the corresponding map $\zeta\eta : t \rightarrow \hat{t}$. It is not difficult to see that $\zeta\eta$ is differentiable, with ζ' converging to 1 as $\eta \rightarrow$. A direct calculation yields the following expression for the directional derivative $T(x')$: of in the direction of $T(x')$:

$$(\nabla' T)\, T = \frac{\kappa_\Gamma(t)}{1 - s\kappa_\Gamma(t)} N(t),$$

where (t, s) is such that $x' = \hat{W}(t) + sN(t)$. In particular, on the set $\omega[\hat{W}]\backslash\omega\eta[\hat{W}]$ we have

$$\kappa_{\Gamma_\eta} \circ \zeta_\eta(t) = \frac{\kappa_\Gamma(t)}{1 - \left(\Gamma_\eta \circ \zeta_\eta(t) - \Gamma(t)\right) \cdot N(t)\, \kappa_\Gamma(t)}, \quad \kappa_{\Gamma_\eta} \circ \zeta_\eta(t) \rightarrow \kappa_\Gamma(t) \quad \text{as } \eta \rightarrow 0. \tag{61}$$

Now we can construct a new surface on $\omega[\hat{W}\eta]$. To this end we will use the equations analogous to (51) and (52). First, we need to adjust the non-zero principal curvature of the surface κ to compensate the shift of the leading curve $\hat{W}\eta$ along the leading segments, so let

$$\kappa_\eta(t) := \frac{\kappa(t)}{1 - \left(\Gamma_\eta \circ \zeta_\eta(t) - \Gamma(t)\right) \cdot N(t)\, \kappa_\Gamma(t)}. \tag{62}$$

We define the leading curve on the surface by

$$\gamma_\eta(\hat{t}) := \int_0^{\hat{t}} \tau_\eta, \quad \gamma_\eta(0) = \gamma(0),$$

with a unique Darboux frame satisfying the following system of equations:

$$\begin{pmatrix} \tau_\eta'(\hat{t}) \\ \nu_\eta'(\hat{t}) \\ n_\eta'(\hat{t}) \end{pmatrix} = \begin{pmatrix} 0 & \kappa_{\Gamma_\eta}(\hat{t}) & \kappa_\eta \circ \zeta_\eta^{-1}(\hat{t}) \\ -\kappa_{\Gamma_\eta}(\hat{t}) & 0 & 0 \\ -\kappa_\eta \circ \zeta_\eta^{-1}(\hat{t}) & 0 & 0 \end{pmatrix} \begin{pmatrix} \tau_\eta(\hat{t}) \\ \nu_\eta(\hat{t}) \\ n_\eta(\hat{t}) \end{pmatrix}, \tag{63}$$

with the initial conditions $\tau\eta(0), \nu\eta(0), n\eta(0) = \tau(0), \nu(0), n(0)$, where ζ^{-1} is the

$$\mathcal{U}_\eta(\hat{t}, s) = \gamma_\eta(\hat{t}) + s\nu_\eta(\hat{t}),$$

for suitable values of $\hat{t}$ and s. Substituting $\hat{t} = \zeta\eta(t)$ we see that $\tau\eta \circ \zeta\eta(t), \nu\eta \circ \zeta\eta(t), n\eta \circ \zeta\eta(t)$

is the solution to the system

$$\frac{d}{dt}\begin{pmatrix} \tau_\eta \circ \zeta_\eta(t) \\ \nu_\eta \circ \zeta_\eta(t) \\ n_\eta \circ \zeta_\eta(t) \end{pmatrix} = \zeta_\eta'(t) \begin{pmatrix} 0 & \kappa_{\Gamma_\eta} \circ \zeta_\eta(t) & \kappa_\eta(t) \\ -\kappa_{\Gamma_\eta} \circ \zeta_\eta(t) & 0 & 0 \\ -\kappa_\eta(t) & 0 & 0 \end{pmatrix} \begin{pmatrix} \tau_\eta \circ \zeta_\eta(t) \\ \nu_\eta \circ \zeta_\eta(t) \\ n_\eta \circ \zeta_\eta(t) \end{pmatrix}. \tag{64}$$

with the initial conditions $\tau\eta \circ \zeta\eta(0), \nu\eta \circ \zeta\eta(0), n\eta \circ \zeta\eta(0) = \tau(0), \nu(0), n(0)$. Due to (61) and the properties of $\zeta\eta$, the coefficients on the right hand side of (64) converge to the coefficients of system (51) in the L^∞-norm. This implies

$$\left\|(\tau_\eta \circ \zeta_\eta, \nu_\eta \circ \zeta_\eta, n_\eta \circ \zeta_\eta) - (\tau, \nu, n)\right\|_{L^\infty(0,l)} \to 0,$$
$$\|\gamma_\eta \circ \zeta_\eta - \gamma\|_{L^\infty(0,l)} \to 0,$$
$$\left\|\mathcal{U}_\eta - u\right\|_{L^\infty(\omega[\Gamma])} \to 0. \tag{65}$$

Analogously we have

$$\nabla'\mathcal{U}_\eta = \left(T_{\eta,1}\tau_\eta - T_{\eta,2}\nu_\eta,\ T_{\eta,2}\tau_\eta + T_{\eta,1}\nu_\eta\right),$$
$$\nabla'^2\mathcal{U}_\eta = \frac{\kappa_\eta \circ \zeta_\eta^{-1}}{1 - \hat{s}\kappa_{\Gamma_\eta}} \begin{pmatrix} T_{\eta,1}^2 n_\eta & T_{\eta,1}T_{\eta,2} n_\eta \\ T_{\eta,1}T_{\eta,2} n_\eta & T_{\eta,2}^2 n_\eta \end{pmatrix},$$

where the terms on the right-hand sides are functions of $\hat{t} \in [0, l_\varepsilon]$, cf. (63). Let us substitute $\hat{t} = \zeta_\eta(t)$ and $\hat{s} = s - (\hat{W}_\eta \circ \zeta_\eta(t) - \hat{W}(t)) \cdot N(t)$ in the above, so that $\hat{W}_\eta(\hat{t}) + \hat{s}\, N_\eta(\hat{t}) = \hat{W}(t) + sN(t)$ everywhere in $\omega[\hat{W}] \backslash \omega_\eta[\hat{W}]$. We obtain, in particular, via (61) and (62) that

$$\frac{\kappa_\eta \circ \zeta_\eta^{-1}(\hat{t})}{1 - \hat{s}\kappa_{\Gamma_\eta}(\hat{t})} = \begin{cases} \dfrac{\kappa(t)}{1 - s\kappa_\Gamma(t)} & \text{on } \omega[\Gamma] \backslash \omega_\eta[\Gamma], \\ \kappa(t) & \text{on } \omega_\eta[\Gamma]. \end{cases}$$

Then from (57), (58) and (65) we derive the following convergence properties:

$$\left\|\nabla'\mathcal{U}_\eta - \nabla' u\right\|_{L^\infty(\omega[\Gamma])} \to 0,$$
$$\left\|\nabla'^2\mathcal{U}_\eta - \nabla'^2 u\right\|_{L^2(\omega[\Gamma])} \to 0,$$
$$\left\|\nabla'^2\mathcal{U}_\eta - \nabla'^2 u\right\|_{L^\infty(\omega[\Gamma] \backslash \omega_\eta[\Gamma])} \to 0, \tag{66}$$
$$\mathrm{II}_\eta = \mathrm{II} \text{ on } \omega[\Gamma] \backslash \omega_\eta[\Gamma], \quad \text{as } \eta \to 0,$$

where II_η is the matrix of the second fundamental form of U_η.

The estimates (59), (66) imply the validity of (56). Indeed, since the number of bodies and arms of u is finite, we can carry out the above construction for every arm (on bodies the surface remains unchanged), and then assemble an $H^2(\omega)$ isometric surface starting from an arbitrary arm or body and attaching sequentially to a current piece the next one by applying a rigid motion, so that the resulting surface satisfies the properties analogous to (66). Then choosing a sufficiently small η and denoting the corresponding surface by $u\delta$ and the union of all $S^{\hat{W}}$ by $S\delta$ yields the estimate (56). Finally, we split each arm of $u\delta$ into smaller arms $\omega[\tilde{W}_r]$ corresponding to the cylindrical parts with rational directions and the remainder.

We conclude the proof with an argument for additional regularity of $u\delta$, which is claimed in the statement. On each arm we can approximate the corresponding curvature κ in the L^2-norm by an infinitely smooth function. We then reconstruct the pieces of the new surface using the Eqs. (51), (52) and $\gamma' = \tau$ with appropriate initial conditions, and attach them together in a fashion similar to the above. Taking sufficiently good approximations of κ we ensure the validity of the estimate (56) for the new surface.

The following scaling argument allows us to obtain the boundedness of $II\delta$ in ω. Due to the smoothness of κ and (54) the matrix of the second fundamental form $II\delta$ is bounded on any compact subset of ω. Without loss of generality we can assume that ω contains the origin. Since ω is convex by the assumption, it is a star-shaped domain, i.e. $\omega\eta := (1 - \eta)\omega \subset \omega$, where we use η again to denote a small positive parameter. Let us define the scaled surface: $u\delta,\eta(x') := (1-\eta)^{-1}u\delta\ (1-\eta)x'$, $x' \in \omega$. It is easy to see that $u\delta,\eta$ is an isometric surface (this is why we need to use the coefficient $(1 - \eta)^{-1}$ in front of $u\delta$). In addition, one has $II\delta,\eta\,(x') = (1 - \eta)II\delta\ (1 - \eta)x'$ and $II\delta,\eta \in L^\infty(\omega)$. Therefore, for any quadratic form Q, in particular for Q^{sc}_{hom}, we obtain

$$\int_{\omega} Q\left(\mathrm{II}_{\delta,\eta}(x')\right) dx' = (1-\eta)^2 \int_{\omega} Q\left(\mathrm{II}_{\delta}((1-\eta)x')\right) dx' = (1-\eta) \int_{\omega_\eta} Q\left(\mathrm{II}_{\delta}(x')\right) dx'.$$

Hence, choosing small enough η and re-denoting $u\delta,\eta$ by $u\delta$, we arrive at (56).

7. LIMIT FUNCTIONAL AND Ŵ-CONVERGENCE IN THE "SUPERCRITICAL" REGIME $H \ll$ E^2

As follows from the compactness result of Sect. 3 and the isometry constraint of Sect. 5, in the case when $h \ll \varepsilon^2$ we should define the homogenised energy functional as

$$E_{\text{hom}}^{\text{sc}}(u) := \frac{1}{12} \int_{\omega} Q_{\text{hom}}^{\text{sc}}(\mathrm{II})\, dx', \tag{67}$$

with the elastic stored-energy function

$$Q_{\text{hom}}^{\text{sc}}(\mathrm{II}) := \min \int_{Y} Q_2\left(y, \mathrm{II} + \nabla_y^2 \psi\right) dy, \quad \psi \in H_{\text{per}}^2(Y), \quad \det\left(\mathrm{II} + \nabla_y^2 \psi\right) = 0. \tag{68}$$

Zero determinant condition is rather restrictive in the case of periodic functions. Next we state a theorem describing the structure of periodic functions ψ subject to this condition, first recalling that II can be represented in the form (54). The coefficient $-\kappa(1 - s\kappa\hat{W})^{-1}$ is one of the two principal curvatures of u, the other principal curvature is always zero.

Theorem 8 *Let* II *be a non-zero matrix of the form* (54) *and let* ψ ∈ *H* 2$_{\text{per}}$ (*Y*) *be such that*

$$\det\left(\mathrm{II} + \nabla_y^2 \psi\right) = 0. \tag{69}$$

Then the derivative of ψ *in the direction* $(-T2, T1)^{\mathsf{T}}$ *is zero, i.e. one has*

$$\psi(y_1, y_2) = \psi_T(T_1 y_1 + T_2 y_2)$$

for some ψ*T* . *In particular, we have* ψ*T* = ψ*T* (*y*1) *or* ψ*T* = ψ*T* (*y*2), *and* ψ*T is* 1*-periodic for T* = $(1,0)^{\mathsf{T}}$, *and T* = $(0,1)^{\mathsf{T}}$, *respectively. For other rational T such that T*1/ *T*2 = *p*/*q* ∈ Q, *where p*/*q is an irreducible fraction, the function* ψ*T is periodic with period P* = |*T*1/ *p*| = |*T*2/*q*|. *For irrational vectors T the function* ψ*T* (*and, hence the function* ψ) *is constant.*

Proof **The proof is based on a result by Pakzad [20]. Let us denote**

$$v(y) := \frac{1}{2} y_1^2 \mathrm{II}_{11} + y_1 y_2 \mathrm{II}_{12} + \frac{1}{2} y_2^2 \mathrm{II}_{22} + \psi,$$

so that $\det(\nabla^2 v) = 0$ is equivalent to Eq. (69). Then ∇v is continuous and for any regular convex domain *D* in R^2 and any *y* ∈ *D* there exists a segment of line passing through *y* and connecting two points on the boundary of *D* such that ∇*y* v = II*y* + ∇*y* ψ is constant along this segment. Since the periodicity and continuity (by the Sobolev Embedding Theorem) of ψ imply its boundedness, it follows that the segment along which ∇*y* v is constant is parallel to the vector $(-T2, T1)^{\mathsf{T}}$: along this direction II*y* is constant. Hence ∇*y* ψ is constant in the direction $(-T2, T1)^{\mathsf{T}}$, and the derivative of ψ in this direction is zero due to the boundedness of ψ. The rest of the statement of the theorem easily follows from the periodicity of ψ.

The above theorem implies that at the points x' where the principal directions of the isometric surface u are irrational we have $Q^{sc}_{hom}(\mathrm{II}) = \int_Y Q_2(y, \mathrm{II})dy$, while on the rational directions the corrector ψ is non-trivial and in general one has $Q^{sc}_{hom}(\mathrm{II}) < \int_Y Q_2(y, \mathrm{II})dy$. This means that in general $Q^{sc}_{hom}(\mathrm{II})$

$$\psi(y_1, y_2) = -\frac{\kappa}{1 - s\kappa_\Gamma}\psi_T(T_1 y_1 + T_2 y_2), \tag{70}$$

where $\psi T(t)$ is the solution of the minimisation problem

$$\min \int_Y Q_2\left(y, \tilde{\mathrm{II}}(1 + \psi_T''(T_1 y_1 + T_2 y_2))\right) dy, \quad \psi_T \in H^2_{per}(0, P), \tag{71}$$

with II defined by (55). Notice that the Sobolev embedding theorem implies that ψT is continuously differentiable.

Theorem 9 (Ŵ-convergence) *Suppose that $h \ll \varepsilon^2$, i.e. $h\varepsilon^{-2} = o(1)$ as $h \to 0$. Then the rescaled sequence of functionals $h^{-2} Eh$ Ŵ-converges to the limit functional E^{sc}_{hom} in the sense that*

1. **Lower bound.** *For every bounded bending energy sequence $uh \in H^1(\ )$ such that $\nabla' uh$* converges to ∇′uu strongly in L2(Ω), u∈H2iso(ω), the following inequality holds:

$$\liminf_{h\to 0} h^{-2} E_h(u_h) \geq E^{sc}_{hom}(u).$$

2. **Recovery of the lower bound.** *For every* $u \in H^2(\omega)$ *there exists a sequence* urech∈H1(Ω) such that ∇′urech converges to ∇′u strongly in L2(Ω) and

$$\lim_{h\to 0} h^{-2} E_h(u_h^{\text{rec}}) = E_{\text{hom}}^{\text{sc}}(u).$$

Proof Lower bound. The proof follows the derivation of the lower bound in the case $\varepsilon \gg h \gg \varepsilon^2$, up to the last inequality in (35), whose right-hand side can have $Q_{\text{hom}}^{\text{sc}}$ replaced $Q_{\text{hom}}^{\text{sc}}$ given by (68), thanks to Theorem 5.

Recovery sequence. A natural method of approximating the deformation of an elastic plate is to first write an expression $u = u(')$ for the mid-surface of the deformed plate, which takes into account the microscopic oscillations in the definition of the homogenised functional (represented by the function ψ), and then "add thickness" to this surface. One possible approach to deriving the mid-surface expression is via an asymptotic expansion, with $u(x')$ as the leading-order term followed by a series of correctors in sequential powers of ε. In the case $h \ll \varepsilon^2$, the first corrector involving the function ψ is shown to satisfy Eq. (69), see also Appendix 2, which is the solvability condition for the equation at the next order in ε. However, higher-order equations all involve two unknown functions and their solution is not straightforward. As was already mentioned in the previous section, in our construction we adopt a different approach based on properties of isometric immersions which we describe next. The details of the formal asymptotic argument are given in Appendix 2.

As was shown in Theorem 8, the corrector ψ is constant whenever *T* is irrational. This implies that if the measure of the set on which

T is rational is zero, then the limit elastic energy is simply

$$\frac{1}{12}\int\limits_{\omega} Q_{\text{hom}}^{\text{sc}}(\text{II})dx' = \frac{1}{12}\int\limits_{\omega\times Y} Q_2(y,\text{II})dydx'.$$

Thus, the only case that remains to be considered is when the above mentioned set has non-zero measure.

Suppose that $u \in H^2$ (ω). By Theorem 7 we can choose a sequence of regularised surfaces $u\delta_j, \delta j \to 0$, such that

$$\begin{aligned} &\left|E_{\text{hom}}^{\text{sc}}(u) - E_{\text{hom}}^{\text{sc}}(u_{\delta_j})\right| \leq \frac{1}{j}, \\ &\int\limits_{S_{\delta_j}}\int\limits_{Y} Q_2(y,\text{II}_{\delta_j})dx'dy \leq \frac{1}{j}. \end{aligned} \tag{72}$$

We first construct an approximating sequence of plate deformations for $u\delta_j$ for each j and then apply a diagonalisation procedure to obtain a recovery sequence for u. Notice that all objects below depend on the parameter δj, however we drop it from the notation in most cases. Let us consider an arm $\omega[\hat{W}r]$ of $u\delta_j$ corresponding to a rational direction T. On this arm the leading curve $\hat{W}r$ is a segment of a straight line, T is constant, $\kappa\hat{W}_r = 0$, the principal curvature is $-\kappa(t)$ (see (54)), and the vector ν of the Darboux frame (τ,ν,n) of $u\delta_j \circ \hat{W}r$ is constant. Let $\psi = -\kappa(t)\psi T\,(T1\,y1 + T2\,y2)$ be the minimiser of (68), cf. also (70). We have

$$\begin{aligned} \text{II} &= -\kappa\begin{pmatrix} T_1^2 & T_1T_2 \\ T_1T_2 & T_2^2 \end{pmatrix}, \\ Q_{\text{hom}}^{\text{sc}}(\text{II}) &= \min\int\limits_{Y} Q_2\left(y,\text{II} - \kappa\nabla_y^2\psi_T(T_1y_1 + T_2y_2)\right)dy \\ &= \min\int\limits_{Y} Q_2\left(y,\text{II}(1+\psi_T''(T_1y_1+T_2y_2))\right)dy. \end{aligned}$$

Let $\psi T,j$ be infinitely smooth functions approximating ψT in $L^2(0, l)$. We will make a more precise choice of these functions later. In order to incorporate the corrector $\psi T, j$ in the equation of the approximating surface, we need to understand the relation between the coordinates (t, s), x' and the underlying εY-periodic lattice. From $x' = \hat{W}(t)+sN = \hat{W}(0)+ tT + sN$ we derive $t = T \cdot (x' - \hat{W}(0))$. Applying the unfolding operator to $\psi T,j(\varepsilon^{-1}t) = \psi T,j(\varepsilon^{-1} T \cdot (x' - \hat{W}(0)))$ we get

$$\mathcal{T}_\varepsilon(\psi_{T,j}(\varepsilon^{-1}t)) = \psi_{T,j}(T \cdot [\varepsilon^{-1}x' - \varepsilon^{-1}\Gamma(0)] + T \cdot y)$$

with $y = \{\varepsilon^{-1} x' - \varepsilon^{-1}\hat{W}(0)\}$, where $[\cdot]$ and $\{\cdot\}$ denote the integer and the fractional parts. Hence, in order to align the argument $\varepsilon^{-1}t$ with the lattice, we need to shift it by adding. Hence, in order to align the argument ε−1tε−1t with the lattice, we need to shift it by adding t∗ε=T·{ε−1Γ(0)}tε∗=T·{ε−1Γ(0)}:

$$\mathcal{T}_\varepsilon(\psi_{T,j}(\varepsilon^{-1}t + t_\varepsilon^*)) = \psi_{T,j}(T \cdot ([\varepsilon^{-1}x'] - [\varepsilon^{-1}\Gamma(0)]) + T \cdot y) = \psi_{T,j}(T \cdot y)$$

with y={ε−1x′}y={ε−1x′}. On the domain $[0, l]$ of ΓrΓr we define a corrected tangent vector

$$\tau_\varepsilon(t) := \tau(t)\sqrt{1 - (\varepsilon\kappa(t)\psi'_{T,j}(\varepsilon^{-1}t + t_\varepsilon^*))^2} + \varepsilon\kappa(t)\psi'_{T,j}(\varepsilon^{-1}t + t_\varepsilon^*)n(t). \tag{73}$$

so that $\tau\varepsilon(t)| = 1$. For small values of ε, the square root in the above is well defined. The approximating leading curve $\gamma\varepsilon$ is given by integration of $\tau\varepsilon$, namely

$$\gamma_\varepsilon(t) := \int_0^t \tau_\varepsilon, \quad \gamma_\varepsilon(0) = \gamma(0).$$

Now we set

$$u_\varepsilon\big(\Gamma(t)+sN\big) := \gamma_\varepsilon(t)+s\nu, \quad \text{whenever} \quad \Gamma(t)+sN \in \omega[\Gamma_r].$$

Let us estimate the difference between u and uεuε. First, it is clear that

$$\tau_\varepsilon = \tau + O(\varepsilon), \quad n_\varepsilon = n + O(\varepsilon), \quad \gamma_\varepsilon = \gamma + O(\varepsilon), \quad \text{as } \varepsilon \to 0, \tag{74}$$

where $n\varepsilon := \tau\varepsilon \wedge \nu$. Here and until the end of this section $o(\varepsilon)$, $O(\varepsilon)$ are understood in terms of the L^∞-norm. Hence, one has

$$\|u_\varepsilon - u_{\delta_j}\|_{L^\infty(\omega[\Gamma_r])} \to 0, \quad \text{as } \varepsilon \to 0.$$

Second, we have (cf. (53))

$$\nabla' u_\varepsilon = (T_1\tau_\varepsilon - T_2\nu \,|\, T_2\tau_\varepsilon + T_1\nu)\,.$$

It follows immediately that

$$\|\nabla' u_\varepsilon - \nabla' u_{\delta_j}\|_{L^\infty(\omega[\Gamma_r])} \to 0. \tag{75}$$

We carry out this procedure for all rational arms $\omega[\hat{W}r]$, leaving the surface unchanged on the rest of ω. We then assemble a new surface denoting it by $u\varepsilon$ in the same way it was done in the proof of Theorem 7 using rigid motions. It is clear from the construction that

$$u_\varepsilon \to u_{\delta_j} \text{ in } W^{1,\infty}(\omega), \quad \text{and, in particular, in } H^1(\omega). \tag{76}$$

Now we "add thickness" to the plate and define the approximating deformation as

$$u_{\delta_j,h} := u_\varepsilon(x') + hx_3 n_\varepsilon(x') + h^2 \frac{x_3^2}{2} d_j(x', \varepsilon^{-1}x'), \tag{77}$$

where $d_j(x', y)$ is some function from $C^\infty(\omega \times Y)$ periodic with respect to y, which will be chosen later and the normal n_ε is defined by $n_\varepsilon = \partial_1 u_\varepsilon \wedge \partial_2 u_\varepsilon$ on non-rational arms and bodies. The gradient of $u_{\delta_j,h}$ is given by

$$\nabla_h u_{\delta_j,h} = (\nabla' u_\varepsilon | n_\varepsilon) + hx_3(\nabla' n_\varepsilon | d_j) + h^2 \frac{x_3^2}{2}(\nabla' d_j + \varepsilon^{-1}\nabla'_y d_j | 0).$$

Clearly, $(\nabla' u_\varepsilon, n_\varepsilon) \in SO(3)$, thus we have

$$h^{-1}\left((\nabla' u_\varepsilon | n_\varepsilon)^\top \nabla_h u_{\delta_j,h} - I_3\right) = x_3(\nabla' u_\varepsilon | n_\varepsilon)^\top (\nabla' n_\varepsilon | d_j) + o(\varepsilon).$$

On each arm $\omega(\hat{W})$ that does not belong to the set of rational arms $\omega(\hat{W}_r)$, the surface u_ε is a rigid motion of u. Such a transformation preserves the matrix of the second fundamental form, i.e. for the matrix of the second fundamental form of u_ε we have $\mathrm{II}_\varepsilon = (\nabla' u_\varepsilon)^\top \nabla' n_\varepsilon = \mathrm{II}$ on $\omega(\hat{W})$.

Let us consider a rational arm $\omega(\hat{W}_r)$. Notice that for κ_ε, which is given via $\tau' = \kappa_\varepsilon n_\varepsilon$, after a simple calculation using (73) and (74) one has

$$\kappa_\varepsilon = \kappa + \kappa \psi''_{T,j}(\varepsilon^{-1}t) + O(\varepsilon). \tag{78}$$

Since $n'_\varepsilon = -\kappa_\varepsilon \tau_\varepsilon$ we have

$$\partial_i n_\varepsilon = -\kappa_\varepsilon \tau_\varepsilon T_i, \quad i = 1, 2.$$

Then from (74), (75) and (78) we obtain

$$(\nabla' u_\varepsilon)^\top \nabla' n_\varepsilon = -(\kappa + \kappa \psi''_{T,j})\widetilde{\mathrm{II}} + O(\varepsilon).$$

We conclude that

$$h^{-1}\left((\nabla' u_\varepsilon, n_\varepsilon)^\top \nabla_h u_{\delta_j,h} - \mathcal{I}_3\right) = \begin{cases} x_3\left(\mathrm{II}|\tilde{d}_j\right) + o(\varepsilon) & \text{if } x' \in \omega_{ir} \cap S_\delta, \\ x_3\left(-(\kappa + \kappa\psi''_{T,j})\widetilde{\mathrm{II}}|\tilde{d}_j\right) + O(\varepsilon) & \text{if } x' \in \omega_r, \end{cases} \tag{79}$$

where $\tilde{d}j := (\nabla' u, n)^\top dj$, and $\omega = \omega r \cup \omega ir \cup S\delta$ is the natural decomposition of ω into the union of rational arms $\omega r = \hat{W}_r\ \omega[\hat{W}r]$, the residual set of rational segments $S\delta$, and the rest of the set ωir. We understand the notation $\psi T, j$ as a family of functions whose members are related to rational arms $\omega[\hat{W}r]$ via the corresponding directions T.

Now we are ready to define a recovery sequence and prove the convergence of the elastic energy. We use the frame indifference of W, its Taylor expansion near I3, the unfolding operator and (79) to get

$$\begin{aligned} &h^{-2}E_h(u_{\delta_j,h}) \\ &= h^{-2}\int_\Omega W(\varepsilon^{-1}x', \nabla u_{\delta_j,h})dx = h^{-2}\int_\Omega W\left(\varepsilon^{-1}x', (\nabla' u_\varepsilon|n_\varepsilon)^\top \nabla_h u_{\delta_j,h}\right)dx \\ &= \frac{1}{12}\int_{\omega_{ir}\cap S_\delta}\int_Y Q_3\left(y, \mathcal{T}_\varepsilon(\mathrm{II}|\tilde{d}_j)\right)dx'dy + \frac{1}{12}\int_{\omega_r}\int_Y Q_3\left(y, \mathcal{T}_\varepsilon(-(\kappa+\kappa\psi''_{T,j})\widetilde{\mathrm{II}}|\tilde{d}_j)\right)dx'dy \\ &\quad + \int_\Omega O(\varepsilon)dx + \frac{1}{12}\int_{\Lambda_\varepsilon} Q_3\left(\varepsilon^{-1}x', h^{-1}((\nabla' u_\varepsilon, n_\varepsilon)^\top \nabla_h u_{\delta_j,h} - \mathcal{I}_3)\right)dx'. \end{aligned}$$

Notice that we can use the Taylor expansion of W due to the fact that $\mathrm{II} \in L^\infty(\omega)$ by Theorem 7. Passing to the limit in the above as $h \to 0$ and taking into account the properties of two-scale convergence we arrive at

$$\lim_{h\to 0} h^{-2} E_h(u_{\delta_j,h}) = \frac{1}{12} \int\limits_{\omega_{ir}\cap S_\delta} \int\limits_{Y} Q_3\left(y, (\mathrm{II}|\tilde{d}_j)\right) dx'dy$$
$$+ \frac{1}{12} \int\limits_{\omega_r} \int\limits_{Y} Q_3\left(y, (-(\kappa + \kappa\psi''_{T,j}(T\cdot y))\widetilde{\mathrm{II}}|\tilde{d}_j)\right) dx'dy.$$

Choosing ψ*T*,*j* and *dj* close enough to the minimisers of the problems (71) and (23) with appropriate arguments we can make the term

$$\frac{1}{12} \int\limits_{\omega_{ir}} \int\limits_{Y} Q_3\left(y, (\mathrm{II}|\tilde{d}_j)\right) dx'dy + \frac{1}{12} \int\limits_{\omega_r} \int\limits_{Y} Q_3\left(y, (-(\kappa + \kappa\psi''_{T,j}(T\cdot y))\widetilde{\mathrm{II}}|\tilde{d}_j)\right) dx'dy$$

arbitrarily close to

$$\frac{1}{12} \int\limits_{\omega_{ir}\cup\omega_r} Q^{\mathrm{sc}}_{\mathrm{hom}}(\mathrm{II}) dx'.$$

For the term corresponding to the residual set SδjSδj we have, in general,

$$\int\limits_{S_{\delta_j}\times Y} Q_3\left(y, (\mathrm{II}|\tilde{d}_j)\right) dx'dy > \int\limits_{S_{\delta_j}} Q^{\mathrm{sc}}_{\mathrm{hom}}(\mathrm{II}) dx',$$

however thanks to the estimate (72) it can be controlled. Thus, for each *j* we can choose

ψ*T*,*j* and *dj* so that

$$\left| \lim_{h\to 0} h^{-2} E_h(u_{\delta_j,h}) - E^{\mathrm{sc}}_{\mathrm{hom}}(u_{\delta_j}) \right| \le \frac{2}{j}.$$

The latter, together with (72), implies that we can choose a subsequence *hj* → 0 so that

$$\left|h_j^{-2} E_{h_j}(u_{\delta_j,h_j}) - E_{\text{hom}}^{\text{sc}}(u)\right| \leq \frac{4}{j}.$$

Passing to the limit as $j \to \infty$ we get the convergence of energies. Thus

$$u_h^{\text{rec}} := u_{\delta_j,h_j}$$

is a recovery sequence. The convergence of the gradients follows from (56), (76) and the construction procedure.

FOOTNOTES

1 The convexity of ωω is a technical requirement which allows us to use the results describing the properties of isometric immersions for the supercritical case $h \ll \varepsilon^2$. From the point of view of the aim of the present work this requirement is insignificant, since the limit effective behaviour of an elastic plate is a local property.

2 We define $H^2_{\text{per}}(Y)$ as the space of functions in $H^2_{\text{loc}}(\mathbb{R}^2)$ that are *Y*-periodic.

3 Note that the expression $r(\varepsilon^{-1}x', \chi_h hG_h) = \chi_h r(\varepsilon^{-1}x', hG_h) + (1 - \chi_h) r(\varepsilon^{-1}x', 0) = \chi_h r(\varepsilon^{-1}x', hG_h)$ is measurable, in view of the facts that $r = r(y, G)$ is a Carathédory function and χhχh is measurable.

4 A direct calculation shows that

$$|\Gamma_n''| \leq |\phi_{n,k}'|(1 - \phi_{n,k}^2)^{-1/2} = 2^{2n}\beta_n|\phi'|(1 - \beta_n^2\phi^2)^{-1/2},$$

which e.g. for the choice $\phi(t) = (\sin(2\pi t))^3$, $t \in [0, 1]$, and $\beta \leq 2\sqrt{2}$ gives the value $C = 4\pi\sqrt{2/3}$.

ACKNOWLEDGMENTS

This work was carried out under the financial support of the Engineering and Physical Sciences Research Council (Grants EP/H028587/1 "Rigorous derivation of moderate and high-contrast nonlinear composite plate theories" and EP/L018802/1 "Mathematical foundations of metamaterials: homogenisation, dissipation and operator theory"). We would like to thank Dr Shane Cooper for helpful remarks.

REFERENCES

1. Acerbi, E., Buttazzo, G., Percivale, D.: A variational definition of the strain energy for an elastic string. J. Elast. **25**(2), 137–148 (1991)

2. Cherdantsev, M., Cherednichenko, K.D.: Two-scale ΓΓ-convergence of integral functionals and its application to homogenisation of nonlinear high-contrast periodic composites. Arch. Ration. Mech. Anal. **204**, 445–478 (2012)

3. Cherdantsev, M., Cherednichenko, K.D., Neukamm, S.: Homogenisation in finite elasticity for composites with a high contrast in the vicinity of rigid-body motions. Preprint (2013) arXiv:1303.1224

4. Ciarlet, P.G.: Mathematical Elasticity: Three-Dimensional Elasticity, vol. I. Elsevier, North-Holland (1988)

5. Cioranescu, D., Damlamian, A., Griso, G.: The periodic unfolding method in homogenization. SIAM J. Math. Anal. **40**(4), 1585–1620 (2008)

6. De Giorgi, E., Franzoni, T.: Su un tipo di convergenza variazionale. Atti Accad. Naz. Lincei Rend. Cl. Sci. Mat. Natur. **58**, 842–850 (1975)

7. Dacorogna, B.: Direct Methods in Calculus of Variations. Springer-Verlag, Berlin (2000)

8. Friesecke, G., James, R.D., Müller, S.: A theorem on geometric rigidity and the derivation of nonlinear plate theory from three-dimensional Elasticity. Commun. Pure Appl. Math. **55**(11), 1461–1506 (2002)

9. Friesecke, G., James, R.D., Müller, S.: A hierarchy of plate models derived from nonlinear elasticity by Gamma-convergence. Arch. Ration. Mech. Anal. **180**(2), 183–236 (2006)

10. Griso, G.: Error estimate and unfolding for periodic homogenization. Asymptot. Anal. **40**(3–4), 269–286 (2004)

11. Hardy, G.H., Littlewood, J.E., Polya, G.: Inequalities, 2nd edn. Cambridge Mathematical Library, Cambridge University Press (1952)

12. Hornung, P.: Approximating W2,2W2,2 isometric immersions. Comptes Rendus Mathematique **346**(3), 189–192 (2008)

13. Hornung, P.: Approximation of flat W2,2W2,2 isometric immersions by smooth ones. Arch. Ration. Mech. Anal. **199**(3), 1015–1067 (2011)

14. Hornung, P., Neukamm, S., Velcic, I.: Derivation of a homogenized nonlinear plate theory from 3d elasticity. Calc. Var. Partial Differ. Equ. **51**(3–4), 677–699 (2014)

15. Le Dret, H., Raoult, A.: The nonlinear membrane model as variational limit of nonlinear three-dimensional elasticity. J. Math. Pures Appl. (9) **74**(6), 549–578 (1995)

16. Müller, S., Pakzad, M.R.: Regularity properties of isometric immersions. Mathematische Zeitschrift**251**(2), 313–331 (2005)

17. Neukamm, S.: Homogenization, linearization and dimension reduction in elasticity with variational methods. Ph.D. Thesis, Munich (2010)

18. Neukamm, S., Olbermann, H.: Homogenization of the nonlinear bending theory for plates. Calc. Var. Partial Differ. Equ. **53**(3–4), 719–753 (2015)

19. Nguetseng, G.: A general convergence result for a functional related to the theory of homogenization. SIAM J. Math. Anal. **20**, 608–629 (1989)

20. Pakzad, M.R.: On the Sobolev space of isometric immersions. J. Differ. Geom. **66**(1), 47–69 (2004)

21. Pantz, O.: Quelques problèmes de modélisation en élasticité non linéaire (Some modeling problems in nonlinear elasticity). These de l'Universite Pierre et Marie Curie (2001)

22. Spivak, M.: A Comprehensive Introduction to Differential Geometry, vol. III. Publish or Perish Inc, Wilmington (1979)

23. Velcic, I.: A note on the derivation of homogenized bending plate model. Preprint (2013) arXiv:1212.2594

24. Jikov, V.V., Kozlov, S.M., Oleinik, O.A.: Homogenisation of Differential Operators and Integral Functionals. Springer-Verlag, Berlin (1994)

25. Zhikov, V.V., Pastukhova, S.E.: On operator estimates for some problems in homogenisation theory. Russ. J. Math. Phys. **12**(4), 515–524 (2005)

Index

R

T